SPACE CHARGE DOMINATED BEAM PHYSICS FOR HEAVY ION FUSION

SPACE CHARGE DOMINATED BEAM PHYSICS FOR HEAVY ION FUSION

Saitama, Japan   December 1998

EDITOR
Yuri K. Batygin
RIKEN, Saitama

American Institute of Physics

AIP CONFERENCE
PROCEEDINGS 480

Woodbury, New York

Editor:

Yuri K. Batygin
RI Beam Factory Project Office
RIKEN
2-1 Hirosawa, Wako-shi
Saitama 351-01
JAPAN

E-mail: batygin@rikaxp.riken.go.jp

The article on pp. 21–30 was authored by a U. S. Government employee and is not covered by the below mentioned copyright.

Authorization to photocopy items for internal or personal use, beyond the free copying permitted under the 1978 U.S. Copyright Law (see statement below), is granted by the American Institute of Physics for users registered with the Copyright Clearance Center (CCC) Transactional Reporting Service, provided that the base fee of $15.00 per copy is paid directly to CCC, 222 Rosewood Drive, Danvers, MA 01923. For those organizations that have been granted a photocopy license by CCC, a separate system of payment has been arranged. The fee code for users of the Transactional Reporting Service is: 1-56396-860-6/99/$15.00.

© 1999 American Institute of Physics

Individual readers of this volume and nonprofit libraries, acting for them, are permitted to make fair use of the material in it, such as copying an article for use in teaching or research. Permission is granted to quote from this volume in scientific work with the customary acknowledgment of the source. To reprint a figure, table, or other excerpt requires the consent of one of the original authors and notification to AIP. Republication or systematic or multiple reproduction of any material in this volume is permitted only under license from AIP. Address inquiries to Office of Rights and Permissions, 500 Sunnyside Boulevard, Woodbury, NY 11797-2999; phone: 516-576-2268; fax: 516-576-2499; e-mail: rights@aip.org.

L.C. Catalog Card No. 99-62728
ISBN 1-56396-860-6
ISSN 0094-243X
DOE CONF- 981211

Printed in the United States of America

Contents

Preface vii
Participants ix
Group Photo xv

TWAC Facility and the Use of the Laser Ion Source for Production of Intense Heavy Ion Beams 1
    B. Sharkov, S. Kondrashev, A. Shumshurov, N. Mescheryakov, I. Rudskoy, S. Homenko, K. Makarov, V. Roerich, A. Stepanov, Y. Satov, H. Haseroth, H. Kugler, N. Lisi, and R. Scrivens

Beam-Beam Compensation Activities at Fermilab. R&D Status 8
    A. Sery, S. Danilov, D. Finley, and V. Shiltsev

An Approach to Fundamental Study of Beam Loss Minimization 21
    R. A. Jameson

Numerical Code for Monte-Carlo Simulation of Ion Storage 31
    N. Alekseev, A. Bolshakov, E. Mustafin, and P. Zenkevich

The Injection and Storage Schemes for Heavy Ion Beams 42
    I. Meshkov, E. Syresin, T. Katayama, and Y. Yano

Electron Cooling with Circulating Electron Beam in GeV Energy Range ... 65
    I. Meshkov

Transverse Electron-Ion Instability in Ion Storage Rings with High Current 74
    P. R. Zenkevich

High Current Induction Linacs at JINR and Perspective of Their Application for Acceleration of Ions 85
    G. V. Dolbilov

Powerful Nanosecond Pulsed Generators for Linear Induction Accelerators at JINR 99
    G. V. Dolbilov, A. A. Fateev, V. A. Petrov, and A. I. Sidorov

Inhomogeneity Smoothing Using Density Valley Formed by Ion Beam Deposition in ICF Fuel Pellet 108
    K. Fujita, T. Kikuchi, D. Takahashi, M. Yazawa, and S. Kawata

Intense-Heavy-Ion-Beam Transport Through an Insulator Beam Guide for Heavy Ion Fusion 120
    T. Kikuchi, S. Kawata, S. Kato, S. Hanamori, and M. Yazawa

Self-Consistent Beam Distribution in Continuous Transport Lines and RF Field 134
    Y. K. Batygin

Particle Dynamics in a DTL for High Intensity Heavy Ion Beams for Inertial Fusion 159
    G. Parisi, H. Deitinghoff, K. Bongardt, and M. Pabst

Halo Formation in Anisotropic Beams 174
    M. Ikegami

Numerical Simulation of Multicomponent Ion Beams in Transport Lines ... 189
    V. Alexandrov, Y. Batygin, N. Kazarinov, V. Shevtsov, and G. Shirkov

Discussion 209
Author Index 217

PREFACE

The realization of a heavy ion fusion program requires delivery of an enormously large powerful beam (10 GeV, total current 50 kA) to a small target (several mm) during a period of approximately 10 nanoseconds by a heavy ion particle accelerator. Achievement of fusion parameters is one of the most challenging problems for accelerator physics. Due to natural repulsion of charged particles via Coulomb forces, beam space charge effects remain the key problem for designers of high intensity accelerators for heavy ion fusion. The subject of this RIKEN Symposium was to review the present understanding of space charge phenomena in high intensity accelerators for HIF and to discuss possible solutions of unresolved problems.

The Workshop was organized and supported by the RIKEN Radioactive Isotope Beam Factory Project Office. The Program Committee consisted of

T. Katayama	(Univ. Tokyo/RIKEN)
Y. Yano	(RIKEN)
R. Jameson	(LANL)
I. Hofmann	(GSI)
R. Davidson	(PPPL)
E. Lee	(LBL)
A. Friedman	(LLNL)
B. Sharkov	(ITEP)
Y. Batygin	(RIKEN)

The Local Organizing Committee included:

A. Yamashita	(RIKEN)
M. Imanishi	(RIKEN)
R. Kuwana	(RIKEN)
T. Nakamura	(RIKEN)
N. Inabe	(RIKEN)
M. Wakasugi	(RIKEN)

The Symposium was attended by 50 scientists. Over a period of three days, many novel ideas in the following subjects were reported and discussed:

Heavy ion fusion projects
High intensity particle sources
Space charge dominated beam transport
Mechanisms of beam emittance growth and halo formation
Transverse and longitudinal beam equilibria
Space charge in recirculators and storage rings
Beam bunching, cooling and stacking
Beam-beam effects
Beam simulation codes

The proceedings of the Workshop will hopefully be useful for beam physicists and accelerator designers of intense high brightness heavy ion accelerators who are working on heavy ion fusion and other related projects.

Yuri Batygin
RIKEN, Saitama, Japan

Participants

John Barnard
   Lawrence Livermore National Laboratory
   PO Box 808, L-440,
   Livermore, CA 94550,
   USA
   barnard@hif.llnl.gov
Yuri Batygin
   RIKEN, 2-1 Hirosawa, Wako-shi, Saitama 351-01,
   Japan
   batygin@rikaxp.riken.go.jp
Julien Bergoz
   BERGOZ Instrumentation
   01170 Crozet
   France
   bergoz@bergoz.com
Michael Craddock
   TRIUMF
   4004 Wesbrook Mall,
   Vancouver, B.C., V6T 2A3
   Canada
   craddock@triumf.ca
Ronald Davidson
   Princeton Plasma Physics Laboratory,
   Princeton University,
   New Jersey, 08543
   USA
   rdavidson@pppl.gov
Gennady Dolbilov
   Joint Institute for Nuclear Research,
   Particle Physics Laboratory,
   141980, Dubna, Moscow Region
   Russia
   dol@sunse.jinr.ru
Anatoly Fateev
   Joint Institute for Nuclear Research,
   Particle Physics Laboratory
   141980, Dubna, Moscow Region
   Russia
   fateev@sunse.jinr.ru
M. Fukuda
   JAERI
   1233 Watanuki
   Takasaki-shi, Gunma 370-12,
   Japan
   fukuda@taka.jaeri.go.jp

Kaoru Fujita
    Department of Quantum Engineering & Systems Science
    University of Tokyo
    Hongo 7-3-1, Bunkyo-ku
    Tokyo 113-8656
    Japan
    fujita@sophie.q.t.u-tokyo.ac.jp

Kazuhiro Fujita
    Dept of Electrical Eng.,
    Nagaoka Univ. of Technology
    Nagaoka, 940-2188
    Japan
    kazz@stn.nagaokaut.ac.jp

Irving Haber
    Naval Research Laboratory
    Code 6790, Washington DC 20375
    USA
    haber@ppd.nrl.navy.mil

Toshiyuki Hattori
    Tokyo Institute of Technology,
    RINR TIT
    2-12-1 Ohokayama
    Meguro-ku,Tokyo 152,
    Japan
    thattori@nr.titech.ac.jp

Enrique Henestroza
    Lawrence Berkeley National Laboratory,
    1 Cyclotron Road
    Berkeley, CA 94720
    USA
    EHenestroza@lbl.gov

Ingo Hofmann
    Gesellschaft f. Schwerionenforschung mbH
    Postfach 110 552, Planckstrasse 1
    64291 Darmstadt 11
    Germany
    I.Hofmann@gsi.de

Masanori Ikegami
    Japan Atomic Energy Research Institute,
    Tokai-mura, Naka-gun, Ibaraki-ken, 319-1195
    Japan
    ikegami@linac.tokai.jaeri.go.jp

Naohito Inabe
    RIKEN, 2-1 Hirosawa, Wako-shi, Saitama 351-01,
    Japan
    inabe@rikaxp.riken.go.jp

Sachiko Ito
    RIKEN, 2-1 Hirosawa, Wako-shi, Saitama 351-01,
    Japan
    ito@rikaxp.riken.go.jp

Robert Jameson
    LANSCE-1 MS H8108
    Los Alamos National Laboratory
    Los Alamos, NM 87545
    USA
    rjameson@lanl.gov
Osamu Kamigaito
    RIKEN, 2-1 Hirosawa, Wako-shi, Saitama 351-01,
    Japan
    kamigaito@rikaxp.riken.go.jp
M. Kanazawa
    NIRS
    4-9-1 Watanuki
    Inage-ku, Chiba-shi, Chiba 263,
    Japan
    kanazawa@nirs.go.jp
Takeshi Katayama
    Center for Nuclear Study, School of Science, University of Tokyo,
    3-2-1, Midoricho, Tanashi, Tokyo 188,
    Japan
    katayama@insac8.cns.s.u-tokyo.ac.jp
Shigeo Kawata
    Nagaoka Univ. of Technology,
    Nagaoka, 940-2188
    Japan
    kawata@nagaokaut.ac.jp
Takashi Kikuchi
    Dept of Electrical Eng.,
    Nagaoka Univ. of Technology
    Nagaoka, 940-2188
    Japan
    kiktak@stn.nagaokaut.ac.jp
Jong-Won Kim
    RIKEN, 2-1 Hirosawa, Wako-shi, Saitama 351-01,
    Japan
    jwkim@rikaxp.riken.go.jp
Misaki Kobayashi
    RIKEN, 2-1 Hirosawa, Wako-shi, Saitama 351-01,
    Japan
    misaki@postman.riken.go.jp
Dmitrii Koshkarev
    Institute for Theoretical and Experimental Physics
    Bolshaja Cheremushkinskaya, 25
    Moscow 117259
    Russia
    koshkarev@vxitep.itep.ru
Shinji Machida
    KEK Tanashi
    3-2-1 Midori-cho, Tanashi, Tokyo, 188-8501
    Japan
    shinji.machida@kek.jp

Shinjiro Matsui
    Tokyo Institute of Technology,
    RINR TIT
    2-12-1 Ohokayama
    Meguro-ku, Tokyo 152
    Japan
Igor Meshkov
    Joint Institute for Nuclear Research,
    141980, Dubna, Moscow region
    Russia
    meshkov@nu.jinr.ru
Edil Mustafin
    Institute for Theoretical and Experimental Physics
    Bolshaja Cheremushkinskaya, 25
    Moscow, 117259
    Russia
    mustafin@vxitep.itep.ru
Yoshiharu Mori
    KEK, 3-2-1, Midori-cho, Tanashi-shi, Tokyo, 188-0002
    Japan
    moriy@mail.kek.jp
Fumiyoshi Nishiguchi
    RIKEN, 2-1 Hirosawa, Wako-shi, Saitama 351-01,
    Japan
    nishiguchi@postman.riken.go.jp
Masao Ogawa
    Tokyo Institute of Technology,
    2-12-1 Oh-okayama
    Nuclear Reactor Dept.
    Meguro-ku, Tokyo
    Japan
    mogawa@nr.titech.ac.jp
Hiromi Okamoto
    Department of Quantum Matter,
    Graduate School of Advanced Sciences of Matter,
    Hiroshima University,
    1-3-1 Kagamiyama, Higashi-Hiroshima 739-8526,
    Japan
    okamoto@kyticr.kuicr.kyoto-u.ac.jp
Giovanni Parisi
    I.A.P. - Frankfurt University
    Robert -Mayer-Str. 2-4,
    D-60054 Frankfurt
    Germany
    parisi@mikro1.physik.uni-frankfurt.de
Takahito Rizawa
    Toshiba Co.
    4-1 Ukishisima-cho
    Kawasaki-ku, Kawasaki-shi
    Kanagawa, 210,
    Japan

Kimikazn Sasa
: TIT, RINR TIT
: 2-12-1 Ohokayama
: Meguro-ku, Tokyo 152,
: Japan

Peter Seidl
: Lawrence Berkeley National Laboratory
: 47-112, 1 Cyclotron Road
: Berkeley, CA 94720
: USA
: PASeidl@lbl.gov

Andrey Sery
: Fermilab, MS 221, PO Box 500,
: Batavia, IL, 60510
: USA
: sery@fnal.gov

Boris Sharkov
: Institute for Theoretical and Experimental Physics
: Bolshaja Cheremushkinskaya, 25
: Moscow 117259
: Russia
: sharkov@vitep5.itep.ru

Grigori Shirkov
: Joint Institute for Nuclear Research,
: Dubna, Moscow region, 141980
: Russia
: shirkov@sunse.jinr.ru

Evgenii Syresin
: Joint Institute for Nuclear Research, LNP,
: Dubna, 141980 Moscow Region,
: Russia
: syresin@nusun.jinr.ru

Masao Takanaka
: RIKEN, 2-1 Hirosawa, Wako-shi, Saitama 351-01,
: Japan
: takanaka@rikaxp.riken.go.jp

T. Uesugi
: KEK Tanashi
: 3-2-1 Midori-cho, Tanashi, Tokyo, 188-8501,
: Japan

Masanori Wakasugi
: RIKEN, 2-1 Hirosawa, Wako-shi, Saitama 351-01,
: Japan
: wakasugi@rikaxp.riken.go.jp

Meigin Xiao
: RIKEN, 2-1 Hirosawa, Wako-shi, Saitama 351-01,
: Japan
: xiao@rikaxp.riken.go.jp

Yasushige Yano
: RIKEN, 2-1 Hirosawa, Wako-shi, Saitama 351-01,
: Japan
: yano@rikaxp.riken.go.jp

Pavel Zenkevich
Institute for Theoretical and Experimental Physics
117259, Bolshaya Cheremushkinskaya ulitsa, 25,
Moscow, Russia
zenkevich@vitep3.itep.ru

RIKEN Symposium on Space Charge Dominated Beam Physics for Heavy Ion Fusion   December 10-12, 1998

TWAC Facility and the Use of the Laser Ion Source for Production of Intense Heavy Ion Beams

Boris Sharkov

ITEP, Moscow, Russia

in collaboration with

S.Kondrashev, A.Shumshurov, N.Mescheryakov, I.Rudskoy

ITEP, Moscow, Russia

S.Homenko, K.Makarov, V.Roerich, A.Stepanov, Yu.Satov

TRINITI, Troitsk, Moscow region, Russia,

H.Haseroth, H.Kugler, N.Lisi and R.Scrivens

CERN, Geneva, Switzerland

Abstract. Current activities on upgrading of ITEP heavy ion accelerator complex in frame of ITEP-TWAC Project are reported. The project being in progress since 1997 is aiming at production of intense (100 kJ/100 ns) heavy ion beams. Basic idea of the project is the application of Non-Liouvillian technique in an existing accelerator facility based on a heavy ion synchrotron for its adaptation to Heavy Ion Fusion related experiments. Special attention is paid to the results on generation of highly charged medium mass and heavy ions in laser produced plasma. Development of key elements of the laser ion source based on the use of a 100 J repetition rate CO_2-laser for filling of ITEP and CERN accelerator facilities in single turn injection mode is presented.

1.

At ITEP - Moscow a TWAC (Terawatt Accumulator) project is in progress now [1], based on the use of the existing heavy ion accelerator complex, including 13 Tm booster ring, 34 Tm heavy ion synchrotron and 2 MV /3 MHz two-gap heavy ion injector, Fig.1. After pre-acceleration the He-like ions of medium-A (i.e. from C_{12} to Co_{59}) will be injected in the booster ring and accelerated up to 0.7 GeV/u, and then stacked in the storage ring using the Non-Liouvillian technique. A large number of cycles of the 1 Hz rep-rate ion source and repetition of the whole acceleration-accumulation process can provide an increase of the final number of accumulated particles by factor ~1000.

A laser ion source (LIS) has to produce about $5 \cdot 10^{10}$ Co^{25+} ions, which are accelerated in the preinjector I-3 up to 1.6 MV/u and then injected into the 13 Tm booster ring UK. After acceleration up to 0.7 GeV/u, a 250 ns long bunch is transferred in a single turn mode to the synchrotron ring U-10 using a non-Liouvillian (stripping) process - the charge state is changed from Co^{25+} to Co^{27+} by passing through a solid foil of about 5 mg/cm^2.

CP480, *Space Charge Dominated Beam Physics for Heavy Ion Fusion,*
edited by Yuri K. Batygin
© 1999 The American Institute of Physics 1-56396-860-6/99/$15.00

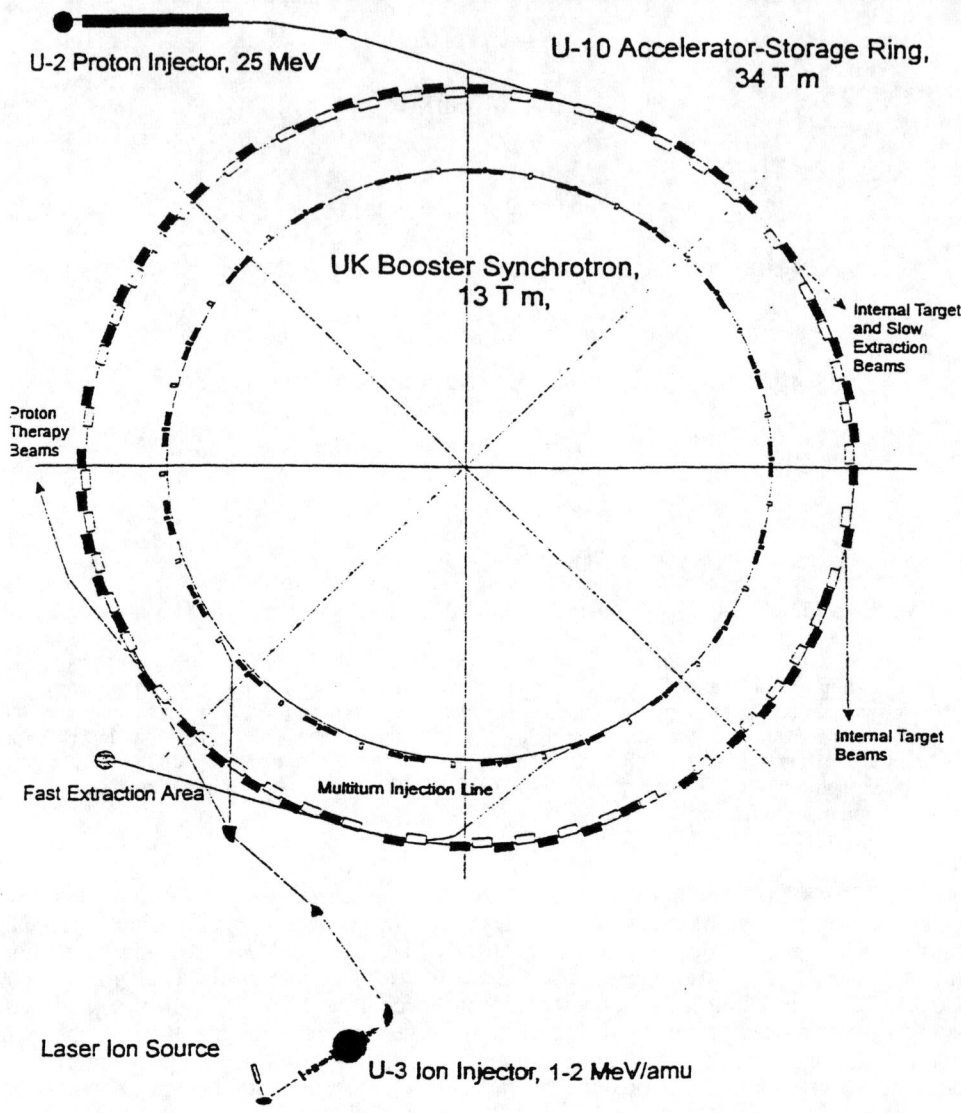

FIGURE 1. Layout of the ITEP-TWAC accelerator complex.

2.

To minimize the affect of the stripping foil, the accumulated beam circulating in the storage ring is directed on the stripper target only during the injection of the next portion of ions coming from the UK booster. For this purpose a beam-bump system, based on the use of two fast coherent deflectors has to be installed. Since the cross section of the accumulated beam is at least 25 times more then the surface of the stripper target, the quality of the beam does not deteriorate significantly from the scattering in the foil. Repetition of this process many times provides accumulation in coasting beam until the Laslett space-charge limit ($\Delta Q \approx 0.16$) is reached in the synchrotron ring with $1.2 \cdot 10^{13}$ ions, corresponding to about 100 kJ stored energy in the beam. A rapid switch on the RF in the synchrotron ring causes ballistic compression (bunch rotation in phase space) of accumulated bunch from a length of 1000 ns down to ~100 ns.

3.

Several effects limiting the final phase space density of accumulated beam are under investigation both analytically (linear theory) and numerically (pic-code). Influence of the space charge tune shift: ($\Delta Q \sim 0.16$) and the development of the longitudinal microwave instability have been found to be not important. However, the transverse coherent instability and the intra beam scattering process have been found to be real limitations for the final phase space density of the beam. As a result a specific momentum spread distribution (80% of ions within $dP/P \leq 4 \cdot 10^{-4}$ and the wings within $dP/P \leq 2 \cdot 10^{-3}$) is suggested. Therefore, an upgraded version of the injector based on the more powerful 10 Hz LIS coupled with the 81 MHz RFQ is foreseen in the second stage of the project.

The expected output parameters of the ITEP-TWAC facility are: the energy on target is 100 kJ in 100 ns, or ~1 TW of beam power. The specific deposition power is ~ 10 TW/g in a spot radius of 500 microns. The temperature reached in a solid Au target is calculated to be up to 40 eV.

Volume energy deposition of heavy ion beam in a target causes generation of a hot strongly compressed matter with extremely high energy density (> 1 MJ/g), producing a uniform intense shock waves in solids and plasmas. A combination of the simple geometry of the energy deposition region (inherent for TWAC beams) with geometrical shape of the target gives a wide variety of different experimental schemes.

The whole scientific program addressed to the ITEP-TWAC facility comprised four main directions. First of all, high energy density in matter physics related to ion fusion. The new facility will give also a big push for experimental investigations in the field of accelerator physic with intense heavy ion beams. Traditionally ITEP is an institute for relativistic nuclear physics: There appears an opportunity to switch from the protons available in ITEP since 30 years to heavy ions. Important is the possibility to accelerate $_{12}C$ and $_{14}N$ ions for tumor therapy.

4.

The most suitable type of ion source is the laser ion source (LIS), capable of producing a sufficient number (10^{10} - 10^{11}) of highly charged ions with atomic mass A ~ 40-60 in 1Hz repetition rate operation mode [2]. The principle of operation of the laser ion source is based on plasma generation by a laser beam focused by a mirror system (or a lens) on a solid movable target. Powerful CO_2 lasers are found to be the best to deliver the energy to evaporate particles from a target, which is made out of the material to be ionized. Production of Helium - like ions of medium-A elements with ionization potentials about ~ 1 keV requires a rather high temperature of plasma [3]. Therefore the development of an efficient LIS is one of the crucial points of the ITEP-TWAC project. A number of candidate ion species like $_{32}S^{14+}$, $_{40}Ca^{18+}$, $_{48}Ti^{22+}$, $_{59}Co^{25+}$ are under consideration now.

There exist a number of requirements to an ion source based on some specific technical features of heavy ion synchrotrons motivating the use of a laser ion source LIS:
(a) pulsed-periodic operation mode with moderate repetition rate ~ 1 Hz;
(b) rather short ion pulse length (~ 3-40 μs) corresponding to the single turn injection mode into the synchrotron ring;
(c) recently sharpened demand for final beam luminosity and related constraints to obtain a maximum possible intensity to fill the synchrotron ring in one turn;
(d) the demand for a wide variety of heavy, multiply charged ion species to be accelerated and used for high energy particle physics experiments.

The LIS meets most of these requirements and has certain advantages compared to other types of heavy ion sources.

An up-dated configuration of a LIS with pre-accelerator will consist of:
- a repetition-rate laser of ~100 J output energy capable to produce more then 10^6 shots without intervention;
- a laser beam transport optical system to uncouple the laser active volume from the light reflected from the target;
- a target ensemble capable of providing 10^5 - 10^6 shots without replacements and interruptions;
- an extraction system extracting the ions out from the expanding laser produced plasma and accelerating them to 10 - 30 keV/u;
- a Low Energy Beam Transport line (LEBT) for matching of the LIS to the subsequent pre-accelerator;
- a pre-accelerator accelerating ions to 0.25 - 1.6 MeV/u (CERN and ITEP options).

The scheme of the 100J laser is shown in Fig. 2. It consists of a master oscillator and a powerful amplifier and a number of spatial filters and saturable cells, determining the high quality of the laser beam [1]. Recent experiments with a mono-pulse prototype of this laser in TRINITI give confidence that such laser can produce the required number of particles per shot.

5.

Development work towards a LIS capable of providing ions for the Large Hadron Collider is in progress at CERN. The LHC filling scheme and the present performance of the accelerator machines lead to the target values at the extraction system of the LIS:

1.5×10^{10} ions of Pb^{25+} in a pulse of 5.5 μs, ε_n(rms) = 0.05 mm mrad, every 1.2 s.

From the start, most of the studies were carried out in close collaboration CERN - ITEP (Moscow) – TRINITI (Troitsk). Joint experimental activities by using laser facilities at TRINITI and CERN and diagnostic equipment from all institutes, as well as the numerical simulations of ITEP and TRINITI teams, resulted in some important achievements:
(1) scaling laws for charge state, ion current density, and pulse duration as function of laser power density;
(2) list of specification parameters of the LIS [2].
(3) complete proposal for the final performance of a LIS to meet the requirements of the CERN and ITEP-TWAC Project.

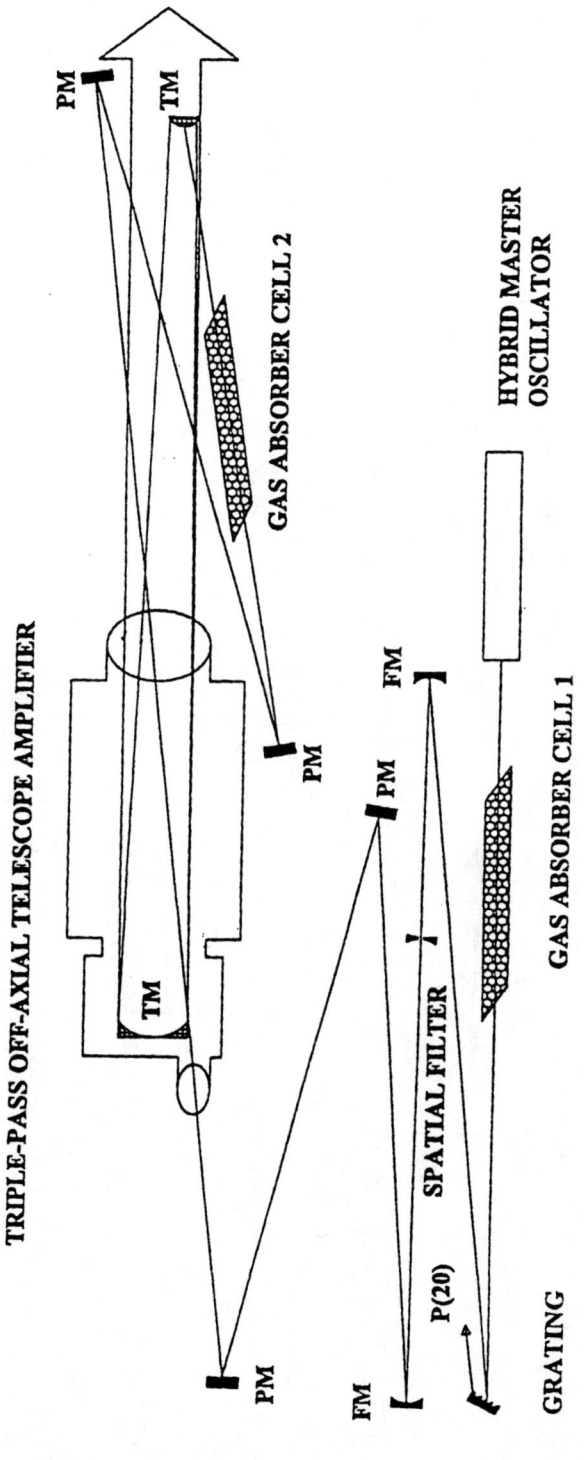

FIGURE 2. Master Oscillator-Power Amplifier scheme of the powerful CO_2 - laser.

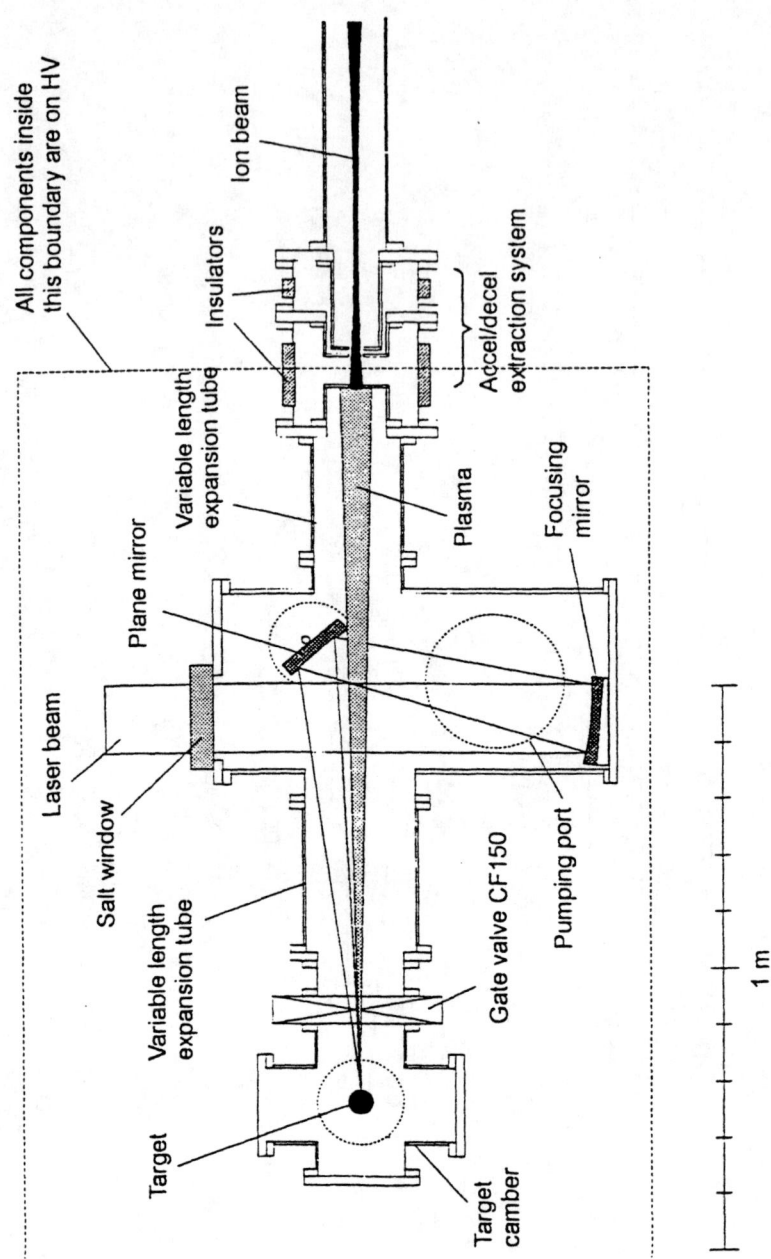

FIGURE 3. Updated design of the laser ion source.

Measured at CERN transmission rate of the LEBT consisting of two solenoids for chosen charge state is about 30 % in a single aperture Faraday cup [4]. Recent analysis of these experiments and simulations with programs like KOBRA3 and PATH, have led to a better understanding of this low transmission. These programs are now the tools in the evaluation of improved transfer schemes.

Big efforts are being invested now in the detailed design of the target positioning mechanism and associated systems: target illumination optical scheme, plasma expansion channel, and extraction system. The updated design, shown in Fig. 3, is based on the target illumination scheme, developed for an interrelated set of parameters.

The target mechanism allows the use of a rotatable cylinder with a surface of more than 400 cm^2, it provides a fine adjustments (+/- 100 µm) of the spot position in two axes, and is capable to accept at least 10^5 shots before replacement. This will allow several days of uninterrupted ion production in a 1 Hz operation mode.

CONCLUSION

Upgrading of the ITEP accelerator complex to a powerful heavy-ion facility promises to provide a very valuable tool for experiments in physics of high energy density in matter, in relativistic nuclear physics, accelerator physics and for medical applications as well.

One of the key elements of the ITEP-TWAC project - the Laser Ion Source - is under intense development now. Experimental results for highly charged ion yield obtained with a mono-pulse prototype of the powerful laser indicate the capability of a 100 J CO_2 laser to provide the required number of ions per laser shot. Much effort is concentrated now on the design of the 1 Hz rep-rate powerful laser amplifier and on an efficient (low particle losses) transport of the extracted ion beam to the subsequent accelerator.

REFERENCES

[1] B.Yu.Sharkov et al., *Nucl.Instr.Meth-A*, 415 (1998) 20-26.
[2] B.Yu.Sharkov et al., *Rev.Sc.Instr.*, V.69(2) (1998) 1035-1039.
[3] I.V.Roudskoy, *Laser Part.Beams* 14, 369, (1996).
[4] P.Fournier et al. Experimental Characterization of Solenoid LEBT for LIS source, CERN PS/HP/Notes 98-14, 99-02.

Beam-beam Compensation Activities at Fermilab. R&D Status.

Andrey Sery, Slava Danilov[1], Dave Finley and Vladimir Shiltsev

*Fermi National Accelerator Laboratory,
MS 221, PO Box 500, Batavia, IL 60510, USA*

Abstract. The beam-beam interaction in the Tevatron produce the betatron tune spread in each bunch and a bunch-to-bunch tune spread. The tune spread sets limits on bunch intensity and luminosity. The beam-beam effects for antiprotons are usually more severe since the proton bunch population is higher.

The beam-beam effects for antiprotons can in principle be compensated with the use of an electron beam with a corresponding charge density. The status of studies of possibilities of the beam-beam compensation is reviewed in this paper.

INTRODUCTION

Investigation of the new frontiers of the elementary particle physics requires permanent increase of performance of the hadron colliders, which are one of the most powerful instruments for such investigations.

The Table 1 represents parameters of the two planned upgrades (Run II and TEV33) of the $p\bar{p}$ Tevatron collider [1,2]. The luminosity increase is achieved mostly due to increase of the bunch population and the number of bunches. Higher bunch population results in enhanced beam-beam effects, namely in increase of the so called betatron tune shift and tune spread (shown in the Table 1) produced by head-on collisions of the bunches in Interaction Points (IP) as well as due to parasitic collisions.

As a result the particles of the beam will cover larger area on the surface of ν_x and ν_y betatron frequencies and this area may cross the lines of higher order resonances that will lead to enhanced diffusion of the particles, decrease of lifetime, growth of emittance and decrease of the luminosity.

The betatron tune shift and tune spread, if it could be arbitrary controlled, is believed to provide a valuable knob for improving beam lifetime and eventually for the maximization of the collider performance.

---

[1] currently at Oak Ridge National Laboratory, Oak Ridge, TN 37831-8218, USA

TABLE 1. Parameters of the Tevatron upgrades.

Parameter		Run II	TEV33
Beam energy	E_b, GeV	1000	1000
Luminosity	$\mathcal{L}, s^{-1}cm^{-2}$	$2.1 \cdot 10^{32}$	$1.2 \cdot 10^{33}$
No. of bunches $(p,\bar{p})$	N_b	36,36	140,121
Min. bunch spacing	τ, ns	396	132
Protons/Bunch	$N_p/10^{11}$	2.7	2.7
Antiprotons/Bunch	$N_{\bar{p}}/10^{11}$	0.75	0.6
p-emittance rms	ε_{np}, $\pi\mu$m·rad	3.3	3.3
$\bar{p}$-emittance rms	$\varepsilon_{n\bar{p}}$, $\pi\mu$m·rad	2.5	2.5
Number of IPs	N_{IP}	2	2
Interaction focus	β^*, cm	37	37
Crossing half-angle	θ_{IP}, mrad	0	0.14
Bunch length	σ_s, cm	37	37→14
$\bar{p}$-tune shift	$\Delta\nu_{\bar{p}}$	~0.020	~0.015
p-tune shift	$\Delta\nu_p$	0.005	0.007
$\bar{p}$ bunch to bunch tune spread	$\delta\nu_{\bar{p}}$	0.007	0.010

The beam-beam compensation techniques based on the use of the intense electron beam have been proposed [3,4] and are under development now [5–8]. The present paper reviews the current status of this investigations.

NONLINEAR COMPENSATION: "ELECTRON COMPRESSOR".

Let us consider schematically the collision of proton and antiproton bunches at the interaction point (see Figure 1).

The proton bunch can effectively be considered as a lens acting on $\bar{p}$ bunch. This additional lens changes the betatron frequency of the on axis $\bar{p}$ by $\Delta\nu_z(0,0) = +\xi^p$ where $\xi^p \equiv N_p r_p / 4\pi\varepsilon_n$ is the so called beam-beam parameter, N_p is the proton bunch population, r_p is the proton classical radius and ε_n is the normalized transverse emittance of the proton bunch.

Since the charge density ρ of the proton bunch is a Gaussian-like, the focusing force F (see Figure 1) of the equivalent lens is a nonlinear function of the transverse

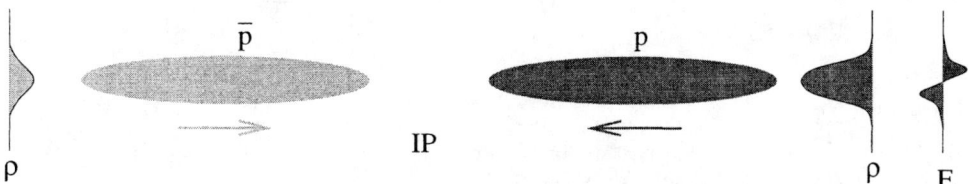

FIGURE 1. Schematics of interaction of round Gaussian beams at the IP.

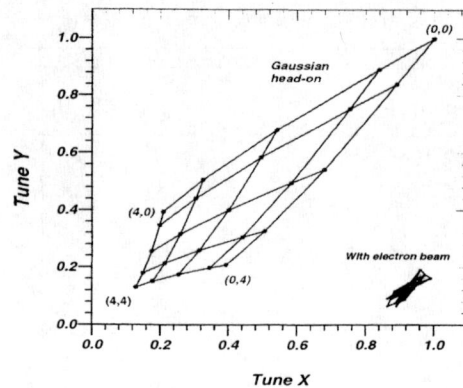

FIGURE 2. Betatron frequencies (tunes) in $\bar{p}$ bunch for particles with different betatron amplitudes (X,Y), head on collision case (large leaf) and the case with compensation by electron beam (small leaf, displaced for clarity) [3]. Tune shift is in units of ξ^p, betatron amplitude is in units of the bunch transverse size σ.

displacement.

Due to nonlinear focusing by p beam the betatron frequencies in $\bar{p}$ bunch are different for particles with different betatron amplitudes (X,Y) as shown on the Figure 2. For the RunII and TeV33 upgrades of Tevatron the spread of betatron frequencies (so called "footprint") of $\bar{p}$ beam is $\Delta\nu_{\bar{p}} \approx 0.02$ that is about the maximum experimentally achieved value for proton colliders. This tune spread $\Delta\nu_{\bar{p}}$ is big enough to cause an increase of particle losses due to higher order lattice resonances.

Compensation of beam-beam induced betatron tune spread within the $\bar{p}$ bunch can be made by an electron beam with equivalent charge distribution [3]. (One should note that usually $N_p \gg N_{\bar{p}}$, so the beam-beam effects for $\bar{p}$ are more severe, that is why we care only about compensation of beam-beam effects for antiprotons.) The schematics of the nonlinear compensation is shown on Figure 3.

The nonlinear focusing of $\bar{p}$ by the proton beam is compensated if a) the electron transverse charge distribution $\rho_e(r)$ is the same as in the proton beam $\rho_p(r)$ (but scaled on r); b) the $\bar{p}$ beam distribution at the "electron compressor" is the same as

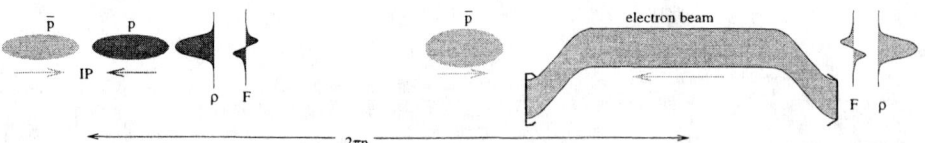

FIGURE 3. Scheme of compensation of the nonlinear beam-beam tune shift in the antiproton bunch by the electron beam with corresponded charge density.

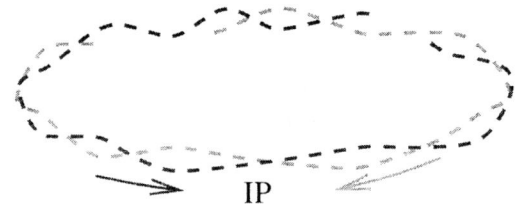

FIGURE 4. Parasitic interactions of proton and antiproton bunches.

at the IP (but scaled on r and with zero dispersion); c) the number of electrons on the path of $\bar{p}$ beam is $N_e = N_p/(1 + \beta_e)$, for example $N_e \approx 4.5 \cdot 10^{11}$ or $J = 1.44$A with $\beta_e = 0.2$ and $L = 3$ m for TEV33.

The electron bunch should have Gaussian transverse distribution in ideal case, however, as it was shown in [3], more realistic and practically more easily achievable distributions can give as good result as the Gaussian one (see Figure 2), the electron beam density in this case was $\propto 1/(1 + (r/\sigma)^8)$ (×-marked line on the Figure 12).

The advanced studies of the nonlinear beam dynamics with the nonlinear focusing by the "electron compressor" have shown that the condition to cancel just the nonlinear tune shift may not be the only condition to satisfy for the antiproton dynamics to be improved. Status of these studies will be reported at the end of the paper.

LINEAR COMPENSATION: "ELECTRON LENS".

Beam beam interactions in colliders with a common vacuum chamber occur not only at the IP but also in hundreds places where the orbits are separated (see Figure 4). These parasitic interactions result in bunch to bunch tune spread. The effect is enhanced by the presence of injection and ejection gaps.

For TeV33 upgrade of the Tevatron the bunch to bunch tune spread is $\Delta \nu_{\bar{p}} \approx 0.01$. Such a tune spread is high enough to enhance dynamic diffusion of particles due to high order resonances, increase background in detectors and limit beam lifetime and luminosity.

The tunes of individual bunches in the $\bar{p}$ beam can be corrected if an additional linear focusing applied to each bunch individually. This focusing can be provided by the field of a wide electron beam ("electron lens") with the current varying on special pattern [4]. The electron lens should be installed in a place where a) electron beam does not interact with proton beam; b) beta-functions $\beta_\perp$ are high enough so the electron current density n_e is reasonable; c) dispersion function is small enough; d) betatron phase advances to IP is close to $2\pi n$.

The possible candidates in Tevatron are the straight section F48 where $\beta_z = 110$m ($\bar{p}$ beam size is $\sigma_z \approx 0.51$mm) and the upstream end of C0 section. Two electron lenses installed in locations with different β_x/β_y are necessary to compensate the x and y bunch-to-bunch tune spread independently (see Figure 5).

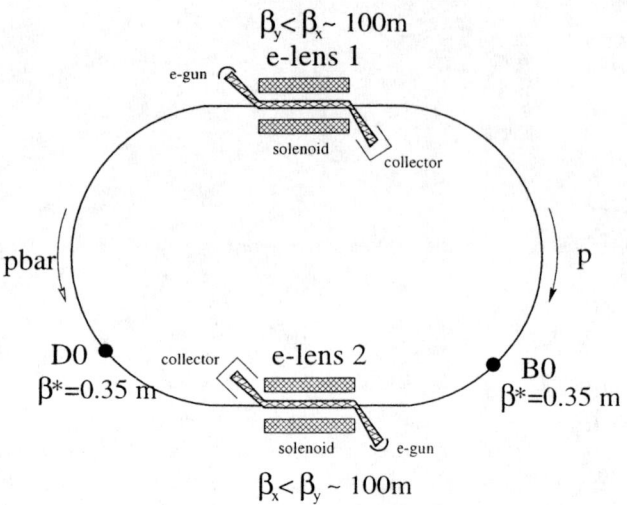

FIGURE 5. Schematics of the two electron lens location in the Tevatron.

For a round, constant density electron beam with total current J, radius a, and interacting with antiprotons over length L, the tune shifts are

$$\xi_\perp^e = -\frac{\beta_\perp}{2}\frac{(1+\beta_e)n_e L r_{\bar{p}}}{\gamma_{\bar{p}}} = -\frac{\beta_\perp}{2\pi}\frac{(1+\beta_e)JLr_{\bar{p}}}{e\beta_e c a^2 \gamma_{\bar{p}}},$$

For example the beam with $J \approx 1.65$ A, $L = 2$ m, $a = 1$ mm, energy 10 kV ($\beta_e = 0.2$) gives $\xi^e \approx -0.01$ in the Tevatron with $\gamma_p \approx 1066$ and beta function $\beta_\perp = 100$m. The electron beam should allow 100% change of the current on 100 ns time scale (corresponds to the distance between bunches) to provide independent influence on different bunches.

Parameters of the electron beam.

The electron beam density n_e is defined from the required tune shift: $\xi_z^e = -\beta_\perp(1+\beta_e)n_e L r_{\bar{p}}/2\gamma_{\bar{p}}$. The length L is defined by the available at Tevatron space $L = 2$m. The electron beam radius a is defined by the $\bar{p}$ beam size. For the electron beam energy the lowest possible value should be chosen provided that a) the current is not limited by the gun itself; b) the electron beam renews faster than the $\bar{p}$-bunch spacing (132 ns).

The gun current is $J = \mathcal{P} \cdot U_a^{3/2}$ where U_a is the anode voltage and $\mathcal{P}$ is the perveance that is typically $\propto 2 \cdot 10^{-6}$ for a diode gun and can be made several times higher for a specially designed gun, such as a convex cathode immersed in a magnetic field [9]. Relying on the gun with perveance $(4-5) \cdot 10^{-6}$, the following optimized parameters of the electron beam can be deduced: the energy 10 kV ($\beta_e = 0.2$), $J \approx 1.65$ A, $L = 2$ m, radius $a = 1$ mm. Such a beam will allow to achieve $\xi^e \approx -0.01$ in Tevatron.

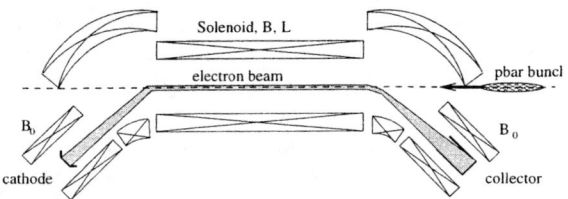

FIGURE 6. A possible layout of the "electron lens".

To decrease the current density to an achievable for oxide cathodes values, one need to use an adiabatic magnetic compression, in which the beam is produced on the cathode with a larger radius a_c in a weak field B_c and then follows the magnetic lines to the region of stronger field B. For the electron lens with cathode current density $2\,A/cm^2$ and cathode radius $a_c = 5$ mm the ratio $B/B_c \equiv a_c^2/a^2$ is to be about 25.

A possible layout of the "electron lens" is shown on the Figure 6.

Experimental test facility at Fermilab.

An experimental installation that should demonstrate the feasibility of the electron lens is now under construction at Fermilab (see Figure 7). This set-up will serve as a prototype of the device that can later be inserted into the Tevatron ring. The test facility is developing in close collaboration of several institutions worldwide, including the Budker Institute of Nuclear Physics (Novosibirsk, Russia) and the INFN LNL (Legnaro, Italy).

The parameters of the experimental installation are about the same as for the full scale device, except a lower magnetic field and current density. The goals of the set-up are a) to obtain 10 kV 2-meter long electron beam with total current up

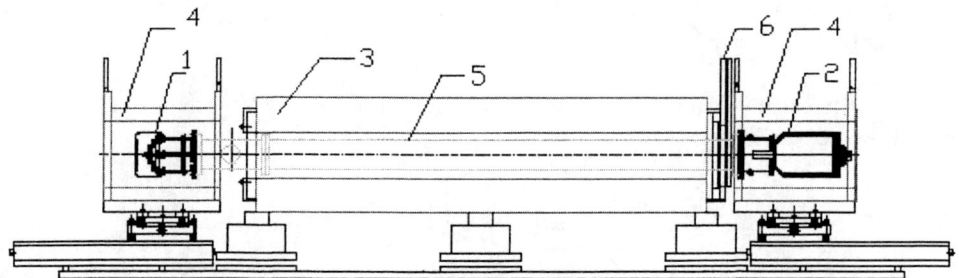

FIGURE 7. Layout of the experimental test facility at Fermilab. 1- electron gun, 10 kV, 2 A max, beam radius at the cathode 5mm, with special near cathode control electrode to change the beam profile; 2- electron collector with electrodes for current distribution analysis; 3- main solenoid, 2 meters long, 4 kG max; 4- additional solenoids, 3.5 kG max; 5- vacuum tube with beam diagnostics; 6- current input/output for solenoids.

to 2A propagating in a precise solenoid magnet; b) to test the current modulation in a few MHz bandwidth; c) to study the beam behavior and to develop necessary beam diagnostics; d) to find the physical and technical solutions needed to build the electron lens for beam-beam compensation in the Tevatron.

The status of the test facility at the end of 1998 is the following. The test facility is assembled, the measurements of the straightness of the magnetic field have been performed. The measured deviation of the magnetic field from the straight line is found to be about $1.6 \cdot 10^{-4}$ rad rms. The field deviation has been measured optically, using a magnetic arrow attached to the mirror which has two rotational degrees of freedom. The gun, collector and vacuum chamber have been installed and the total current of 2 A has been achieved. Investigations of the beam profile, tests of the beam current modulation, and other experiments are under way.

PARASITIC EFFECTS DUE TO ELECTRON BEAM.

The considered ideas of the beam-beam compensation is not a first attempt in history of colliders. One of the first of such attempts was the idea of 4-beam neutralized collisions in e^+e^- e^+e^- colliding rings [10]. The experiment on the dedicated DCI rings has shown, however, that in spite of the significant charge compensation (5-10 times) and increase of the beam-beam parameter ξ from 0.018 to 0.024, no luminosity increase was achieved that was associated with coherent instabilities in the beams [11]. Another attempt was the idea to use neutralized 4-beam collisions in linear colliders, which, as it was shown in corresponded studies [12], also does not give significant benefits due to coherent charge separation instability if $\xi \gtrsim 1$.

Initial proposals of "electron compressor" for nonlinear compensation of the beam-beam induced betatron spread [3] and of the "electron lens" for linear compensation of bunch to bunch tune spread [4] have been exposed to intensive studies of the possible accompanying harmful effects [5–8]. In spite of several possibly harmful effects, which have been found, the idea of beam beam compensation by a single pass electron beam is very attractive. The magnitude of the harmful effects can be made sufficiently small by the proper choice of parameters of the compensating electron beam.

The results of investigations of the most important effects are reviewed briefly in the following sections.

Head tail in $\bar{p}$ beam due to electron beam [7].

Off center collision of the $\bar{p}$ bunch with electron beam results in drift of electrons in crossed magnetic and electrical fields $\vec{B} \times \vec{E}$ so that the head of the $\bar{p}$ sees E_y while the tail will also see E_x (see Figure 8). The resulting change of the betatron amplitude of the tail $\bar{p}$ particle is $\delta x_2 \propto y_1 \xi_x N_{\bar{p}} e/(Ba^2)$.

Taking into account that the head and the tail exchange their position with synchrotron frequency ν_s, one can see that as a result of such a skew interaction

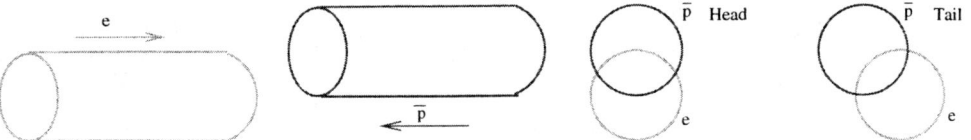

FIGURE 8. Off center collision of the $\bar{p}$ bunch with electron beam.

the horizontal betatron motion (ν_x), the vertical betatron motion (ν_y) and the synchrotron motion (ν_s) become coupled.

The skew coupling of X and Y betatron motion due to the electron beam changes the frequencies of unperturbed transverse motion. The effect is maximal for a pair of closest harmonics ($\nu_x + m\nu_s$) and ($\nu_y + n\nu_s$) (where m, n are integer). With increasing the electron beam current the real parts of their frequencies will become closer and finally collapse, at the same time the imaginary parts will appear, resulting in instability (see Figure 9).

The threshold of this Transverse Mode Coupling Instability (TMCI) in terms of magnetic field was found to be

$$B_{thr} \approx 1.3 \frac{eN_{\bar{p}}\sqrt{\xi_x^e \xi_y^e}}{a^2\sqrt{\Delta\nu\nu_s}}.$$

where $\Delta\nu = \nu_x - \nu_y$. This analytical result was confirmed by numerical simulations. Under the design parameters the minimum magnetic field that will keep the $\bar{p}$ beam stable is $B_{thr} = 17.5$ kG.

Electron beam distortion by elliptical $\bar{p}$ beam [6].

If the set-up will be located at the place where $\beta_x \neq \beta_y$ then axial symmetry is not conserved. The electron beam becomes a rotated ellipse to the moment when the tail of antiproton bunch passes it through, while the head of the bunch sees originally undisturbed round electron beam.

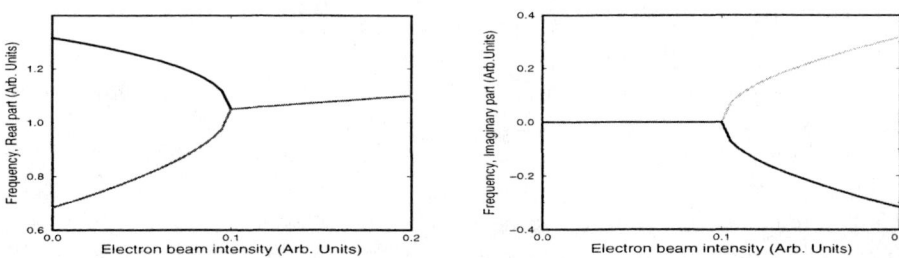

FIGURE 9. Illustration to the mode coupling instability. Frequencies of the antiproton bunch oscillation modes versus the electron beam intensity, Real part (left picture) and Imaginary part (right picture).

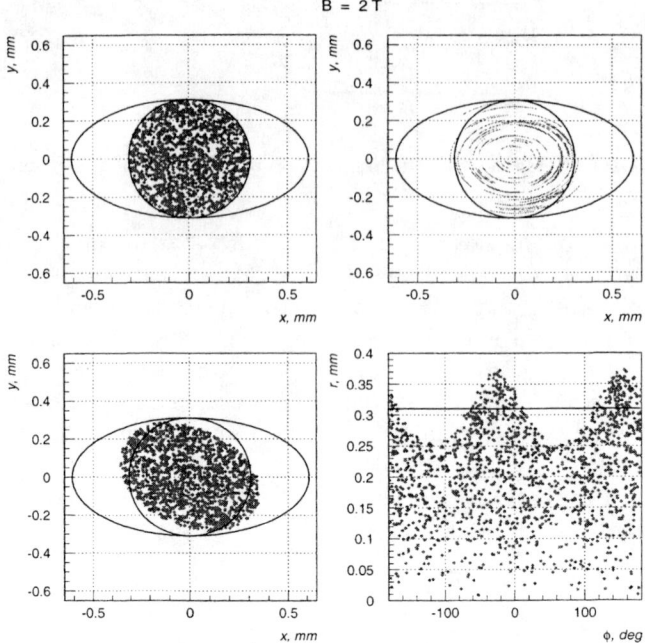

FIGURE 10. "Small" electron beam distortion due to $\bar{p}$ bunch. Round electron beam (radius $a_e = 0.31$mm) interacts with elliptical $\bar{p}$ beam in 2 Tesla solenoid field. Initial distribution (top left), electron velocities (top right), final transverse (bottom left) and azimuthal distributions [6].

The electric fields of the distorted electron beam produce $x-y$ coupling of vertical and horizontal betatron oscillations in the $\bar{p}$ beam. The distortion performs two variations over azimuth $\delta\rho \propto xy \sim \sin(2\theta)$ and the maximum distortion scales as $\delta\rho^{max}/\rho_0^{max} \sim 0.2eN_{\bar{p}}/(a_e^2 B)$.

An example of the distorted electron beam is shown on the Figure 10. Note that distortion appears mostly at the edge of the beam that suggests (and was confirmed analytically and in simulations) that the coupling due to elliptically distorted electron beam can be additionally suppressed if the electron beam has a radius larger than the antiproton beam radius, $a \geq \sigma_{\bar{p}}$.

Field of the elliptic electron beam lead to $x - y$ coupling of betatron oscillations in the $\bar{p}$ beam. The average coupling can be corrected in the Tevatron, however the spread in coupling has to be small enough in order not to affect the $\bar{p}$ beam dynamics.

The high magnetic field can decrease coupling to an acceptable value. If $B = 2$T, the maximum coupling spread is $|\kappa| \simeq 4 \cdot 10^{-4}$ for thin electron beam, and $7 \cdot 10^{-5}$ for wider electron beam. These values are rather small with respect to the typical residual coupling in Tevatron (about 0.001).

$\bar{p}$ emittance growth due to variations of the electron beam [4].

Fluctuations of the electron current $\Delta J/J$ from turn to turn cause time variable quadrupole kicks which lead to a transverse emittance growth of the antiproton bunches. The emittance growth time is more than 10 hours if the to peak-to-peak current fluctuations are smaller than $\Delta J/J \approx 1.8 \cdot 10^{-3}$.

Transverse motion of electron beam result in dipole kick and coherent betatron oscillations experienced by antiprotons. After some decoherence time they will result in antiproton emittance growth. The emittance growth time is more than 10 hours if $\delta X \leq 0.14\,\mu m$.

Deviation of solenoidal magnetic field $\vec{B}$ from a straight line will cause off-center collisions of the antiproton and electron beams. In the case of the non-linear electron lens this may cause unwanted non-linear components of the forces. The effect is small if $\Delta B_\perp / B \lesssim 10^{-4}$.

All these conditions are believed to be achievable.

Nonlinear compensation. Advanced studies [5].

It was thought that the nonlinear compensation will compensate on average the nonlinear focusing of $\bar{p}$ by the proton beam resulting in decreasing the spread of the betatron frequencies (footprint), slowing the dynamic diffusion of particles due to high order resonances, improving radiation background in detectors, enhancing the beam lifetime and luminosity.

The nonlinear compensation, as it was just describe, may not work as desired even assuming that the beam-beam interaction is the only source of nonlinearities.

The proton bunch length expressed in terms of the betatron phase advance is large $\Delta\psi_p \simeq \sigma_s/\beta^* \simeq 1$. In contrast, the electron beam length L is about 2-3 m that gives $\Delta\psi_e \simeq L/\beta \propto 0.01 - 0.02$ (see Figure 11).

Thus, the electron beam kick looks like a delta-function when transformed to the main IP. Consequently, such a short impact from the electrons contains a lot of resonance harmonics, although the average actions due to proton and electron beams are the same. One can reduce the betatron tune spread with a non-linear lens, but this alone does not assure that the motion is more stable than the one

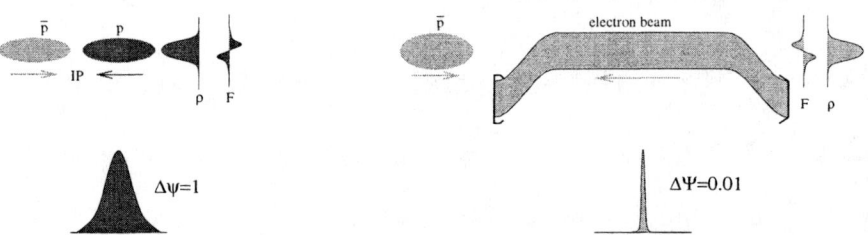

FIGURE 11. Illustration to the nonlinear compensation.

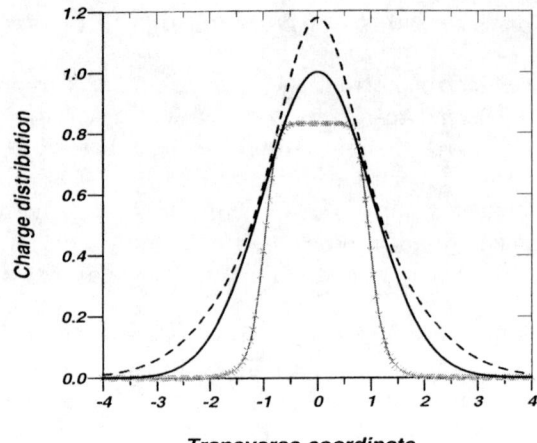

FIGURE 12. Gaussian charge distributions of the protons (solid line), of the electron beam in "electron compressor" (×-marked line) with $\rho_e(r) \propto 0.83/(1 + (r/\sigma)^8)$ and of the optimized electron beam distribution (dashed line) [5].

with no compensation, because the resonance strengths can be more important than the tune spread.

The road to follow is to investigate the possibility to add a single thin nonlinear lens to an arbitrary nonlinear lattice in such a way that the particle motion in the modified structure would become resonance-free, though nonlinear, and at the same time the beam of particles would have a zero footprint.

In axially symmetrical system the existence of such a lens can theoretically be proven. In practical case a numerical method can be used, which consists of minimization of the sum of squared differences of coordinates and momenta at the beginning and at the end of N successive map transformations that will eliminate the Nth order resonance (the frequencies of all particles are equal to the particularly chosen value and the strength of the resonance is equal to zero). This procedure can be done for different N and an optimized electron beam distribution can be found (see an example on the Figure 12).

These investigations should be continued to prove that such a nonlinear insertion indeed improves the beam dynamics.

Stability of the electron beam. Drift instability.

"Drift instability" is the main reason that can limit the beam current in presence of ions.

Ionization of residual gas by electrons produce ions with the rate $dn_i/dt = n_e/\tau_n$ where "time of neutralization" $\tau_n = (\sigma_{\text{ioniz}} v_e n_0)^{-1}$. For our parameters $\tau_n \approx 0.25$ s if the vacuum pressure $P \approx 10^{-9}$ Torr.

Potential well of the electron beam prevents ions to get out of beam in transverse direction. The depth of the well is $U_e = \pi a^2 e n_e (1 + 2 \ln b/a)$ where a, b – radius of electron beam and vacuum chamber, n_e – electron beam density. The ions may also be locked longitudinally if the electron beam is shrunk in the central part of the electron compressor.

Ions should be removed because they a) change charge density, i.e. spoil beam-beam compensation; b) may result in two beam drift instability.

Considering the motion of the charge density centers of the ion and electron beams in dipole approximation, one can find that there is an amplification of a small initial beam separation down along the beam [13–15]. The amplification coefficient is maximal at the resonance frequency

$$K_{\max} = \exp\left(-\frac{\Omega_d L}{v_e \varepsilon''}\right)$$

where $\Omega_d = 2\pi n_e ec/B$ – drift frequency, ε'' – imaginary part of the permeability, in our case approximately $\varepsilon'' \approx n_i/n_e$.

"Drift instability" of electron and ion beams appears when the feedback from the beam end to the beam beginning (e.g. by electrons reflected from the collector) is big enough $K_{\max}\eta > 1$, where η is the feedback coefficient (typically $\eta \approx 10^{-3} - 10^{-4}$).

This mechanism of the instability was confirmed by experiment and the stability condition was found to be

$$\frac{n_i}{n_e} j_e < \frac{v_e^2 B}{4Lc}$$

If $\beta_e = 0.2$, $B = 4$ T, $a = 1$ mm, $I_e = 2$ A, $L = 3$ m then the electron beam is stable if $n_i/n_e < 20$ % (much smaller fraction of ions is allowed from the point of view of beam-beam compensation).

The residual ions are therefore to be cleaned from the electron beam. Special cleaning electrodes will be used for this purpose. The vacuum should also be high enough to ensure that the neutralization time is sufficiently longer than the lifetime of ions in the electron beam. Estimations shows that proper cleaning electrodes together with vacuum better than $P < 3 \cdot 10^{-9}$ Torr will provide acceptable amount of residual ions in the electron beam $n_i/n_e < 0.5 \cdot 10^{-2}$.

CONCLUSION

The beam-beam compensation with an electron beam looks very promising. It provides additional powerful "knobs" to control beam dynamics in the Tevatron collider. No severe requirements on the electron beam were found for the suggested device. We believe that realization of the idea will give benefits for the Tevatron.

REFERENCES

1. J.P.Marriner, FERMILAB-Conf-96/391 (1996); S.D.Holmes, *et.al*, FNAL-TM-1920 (1995).
2. P.Bagley, F.Bieniosek, P.Colestock, *et. al*, FERMILAB-Conf-96/392 (1996).
3. V.Shiltsev, D.Finley, FERMILAB-TM-2008 (1997).
4. V.Shiltsev, FERMILAB-TM-2031 (1997).
5. V.Danilov, V.Shiltsev, FNAL-FN-671 (1998).
6. V.Shiltsev and A.Zinchenko, *Phys. Rev. ST Accel. Beams*, **1**, 064001 (1998).
7. A.Burov, V.Danilov, and V.Shiltsev, FNAL-Pub-98/195 (1998), *Phys. Rev. E*, **59**, No. 3, (1999), in press.
8. V. Shiltsev, V. Danilov, D. Finley, A. Sery, FERMILAB-PUB-98-260 (1998), submitted to *Phys. Rev. ST Accel. Beams*.
9. A.Sharapa, A.Grudiev, D.Myakishev, A.Shemyakin, *Nucl. Instr. Meth. A*, **406** (1998), 169.
10. J.E.Augustin, R.Belbeovh, P.Brunet, *et. al*, in *Proc. 7th Int. Conf. High Energy Accel., Yerevan*, vol.2 (1970), 113.
11. J. Le Duff, M.P.Level, P.C.Martin, E.M.Sommer, H.Zyngier, *11th Int. Conf. High Energy Accel., CERN, Geneva*, Birkhauser Verlag, Basel-Boston-Stuttgart (1980), 707; see also *IEEE Transact. Nucl. Sci.* **26**, No.3 (1979), 3559.
12. N.A.Solyak, Preprint INP 88-44, Novosibirsk (1988); see also in *Proc. 13th Int. Conf. High Energy Accel., Novosibirsk*, vol.1, Budker INP (1986), 151.
13. V.I.Kudelainen, V.Parkhomchuk, D.Pestrikov, *Zh.Tekh.Fiz. (Sov.Phys.-Tech.Phys.)*, **53**, No.5, (1983), 870.
14. A.V.Burov, V.Kudelainen, V.Lebedev, V.Parkhomchuk, A.Sery, V.Shiltsev, Preprint INP 89-116, Novosibirsk (1989, in Russian); Preprint CERN/PS 93-03 (AR), CERN (1993).
15. A.V.Burov, Preprint INP 88-124, Novosibirsk (1988).

An Approach to Fundamental Study of Beam Loss Minimization

R.A. Jameson

LANSCE-1 MS H808
Los Alamos National Laboratory
Los Alamos, New Mexico, 87545 USA

Abstract. The accelerator design rules involving rms matching, developed at CERN in the 1970's, are discussed. An additional rule, for equipartitioning the beam energy among its degrees of freedom, may be added to insure an rms equilibrium condition. If the strong stochasticity threshold is avoided, as it is in realistic accelerator designs, the dynamics is characterized by extremely long transient settling times, making the role of equipartitioning hard to explain. An approach to systematic study using the RFQ accelerator as a simulation testbed is discussed. New methods are available from recent advances in research on complexity, nonlinear dynamics, and chaos.

RMS MATCHING

The top-level specifications for high-intensity particle accelerators include high availability[1,2]. A major availability factor is beam loss minimization, to insure maintainability without remote manipulators[3,4].

There is not much design guidance for beam loss minimization, especially as allowed beam losses are very low, of order 1 nA/m/GeV for rf light ion linacs; e.g., 1 part in 10^8/m/GeV for a 100 mA average current proton linac, resolution beyond the fidelity of existing simulations. The traditional rules are that the beam must be rms matched in the transverse and longitudinal planes:

$$\varepsilon_{tn}^2 = \frac{a^4 \gamma^2 \sigma^{t^2}}{n^2 \lambda^2} = \frac{a^4 \gamma^2}{n^2 \lambda^2} \left(\sigma_o^{t^2} - \frac{I \lambda^3 k (1-ff)}{a^2 (\gamma b) \gamma^2} \right) \tag{1}$$

$$\varepsilon_{ln}^2 = \frac{(\gamma b)^4 \gamma^2 \sigma^{t2}}{n^2 \lambda^2} = \frac{(\gamma b)^4 \gamma^2}{n^2 \lambda^2}\left(\sigma_o^{t2} - \frac{2I\lambda^3 kff}{a^2(\gamma b)\gamma^2}\right) \tag{2}$$

where t and l denote transverse and longitudinal, ε is rms emittance, a and b are transverse and longitudinal rms beam radii (assuming an ellipsoidal distribution), σ is a phase advance with beam current, σ_o is the external field phase advance, λ is rf wavelength, n is the number of $\beta\lambda/2$ in a transverse focusing period, I is the beam current, ff is the geometry factor $\approx a/3b$, γ and β are the relativistic gamma and beta, and $k = \frac{3n^2}{8\Pi}\frac{z_o q 10^{-6}}{mc^2}$. All but two of the parameters must be specified by other means.

The matching equations insure that emittance is not diluted by betatron or synchrotron oscillations. They provide obvious guidance on what to do when there is a transition in the machine of any kind that affects the phase advance: as beam emittance and size would be kept constant across the transition, the focusing per unit length must also be kept constant across the transition. The rms matching equations also guide one in the direction of higher frequency to maximize beam brightness, the ratio of beam current to multidimensional beam emittance, as basically less charge is accelerated per bucket, reducing the space-charge term. However, brightness, characterizing the beam core, is not the same as residual beam loss, which comes from scraping off particles in a tenuous cloud outside the core.

HALO DIAGNOSTICS

Practical designs are done using detailed multiparticle simulation codes to observe *what* happens along a machine, including the effects of various errors. Observed beam boundaries are multiplied by experience-based engineering safety factors to arrive at adequate beam tube dimensions and focusing strengths.

Recently, a great deal has been learned about one common form of practical error - that of rms mismatch[4], which drives parametric resonances. This particular type of error is bounded because a particle's tune is a function of radius. Thus most investigations have concentrated on approximating the boundaries, using a simple model of a stably oscillating core field to influence the motion of test particles arbitrarily initialized[5]. Such a boundary estimation, useful for setting engineering safety factors, emphasizes *what* might happen, if the beam core remains as simple as the model describes, and if particles that describe the larger mismatch orbits could actually reach the appropriate initial conditions.

If one is interested in probing deeper into the mechanisms that can drive particles into unwanted areas of phase-space (e.g., beyond the bore, or into highly nonlinear parts of the rf bucket), it is necessary to go beyond observing *what* happens and ask *how, why, when, where, which,* etc.

In this case, very simplified models and test particle methods are severely limited. The idealized core/test-particle model for rms mismatch, for example, contains no information about particle transport, and can give no information about how, why, which, when, where particles might reach a point in phase-space from which they would make the larger mismatch orbits.

These questions must usually be explored using detailed diagnostics on a fully self-consistent particle simulation code. "Self-consistent" means that all the particles are involved in both producing the field and being influenced by the field. Typically, both the rms characteristics of the beam as well as the particle orbits are functions of time, and this makes simple periodic measurements, such as the usual Poincarè plots, problematic. Methods for observing transport and beam halo are detailed in [4] for a mismatched beam in a self-consistent (x,x',y,y') simulation, symmetrized to (r,r'), in a continuous focusing channel. Particle transport is seen to involve resonance entraining and hopping. Particle escape from the core into the mismatch halo can be precisely defined[a] by defining a local particle tune and noting when the particle's tune accumulates to 2Π in the period of the resonance under observation. This allows very detailed information to be obtained concerning whether particles are in the core or in the halo, when a particle enters or leaves the halo, how long it stays in the halo, and so on.

The single-particle local instantaneous tune in [4] is simply based on the local force balance on the particle, i.e., the local restoring force minus the space-charge force on the particle from all particles inside the particle's physical orbit radius.

Advantages of this method are that it is easy to compute, and rigorously replaces the laborious methods of resonance overlap and turnstile computation that previously were used in the study of nonlinear dynamics and chaos[6].

EQUIPARTITIONING

The two matching equations alone are insufficient to prevent emittance growth of an intense beam as space-charge forces begin to dominate over the emittance terms. For example, steady longitudinal rms emittance growth is seen in conventional designs

---

[a] Most "definitions" of halo are very fuzzy, and there is no consensus.

with fixed accelerating field gradient (which gives decreasing longitudinal focusing with energy) and constant transverse focusing. Also, although rms transverse emittance growth can be kept very small in such designs, the total effective transverse (and longitudinal) emittance growth is large. It would be desirable to add other rules based on space-charge physics to prevent this emittance growth and other sources of potential beam loss.

Only one additional possible rule is presently known - the rule of rms equipartitioning (EP)[7].

An equipartitioned beam bunch has equal energy in the transverse and longitudinal directions, leaving no free, unbalanced energy that could cause emittance growth via coupling mechanisms. A fully matched and equipartitioned bunch would balance out free energy and be matched to all orders, i.e., a truly stationary, equilibrium distribution. Such a distribution is beyond analysis, but such distributions do tend to form in actual accelerator channels. For design purposes, we can place an energy balance condition on the rms beam properties:

$$\frac{e_{ln}\sigma^l}{e_{tn}\sigma^t} = 1 \tag{3}$$

Locally solving Eq. (3) simultaneously with (1) and (2) yields linac designs that are equipartitioned throughout. Addition of the equipartitioning condition puts a strong constraint on the design space; for example, if $\varepsilon_{ln}/\varepsilon_{tn} = 2$ throughout, then $\sigma^l/\sigma^t = 1/2$ throughout.

Thresholds for the low-order coherent instabilities of KV beams have been derived, and shown to be closely related to corresponding instabilities for realistic beam distributions as well[7,8]. In reference [3], it is shown that the EP design trajectory is in a region of tune space that is free of low-order instabilities down to quite large tune depressions, and how the EP condition might be chosen in practical cases.

Simulations of equipartitioned channels show characteristics expected of an equilibrium distribution. For example, consider a channel again having constant accelerating gradient. The transverse focusing is then weakened along the channel to achieve EP. The longitudinal rms emittance growth is no longer seen, and the beam distribution remains "tight", with nearly constant ratio between total and rms beam size (or emittance). The improved longitudinal emittance is important for storage ring injection and has led to acceptance of EP designs for this purpose.

Despite this important evidence, and even acceptance for storage ring injection, almost no further systematic work on this fundamental possibility to achieve an equilibrium beam condition, at least to rms order, has been supported to date.

To some extent, this may be because there are questions about whether EP can provide an equilibrium sufficiently advantageous, for example with respect to total beam size or reduced sensitivity to errors.

One question involves beam size in designs with constant accelerating gradient, an economic constraint in room-temperature linacs. As noted above, this means that the longitudinal focusing decreases with energy. With constant transverse focusing, rms transverse beam size decreases, but total transverse beam size increases considerably. If the transverse focusing is weakened corresponding to the longitudinal focusing to produce EP, the rms transverse beam size naturally grows along the structure. This might seem counterintuitive to beam loss minimization, but total, not rms, beam size is what is important. After acceleration to order 1 GeV, the total beam size for both designs appears, from simulations, to be "roughly similar". Looking only at the output ignores the possibility for lower beam losses all along the structure with the tighter beam distribution of the EP design, compared to the more diffuse distribution with tenuous halo characteristic of the constant focusing design. Systematic studies are needed, including error sensitivity studies.

Another reason is fundamental to the accelerator dynamics in the typical design tune space, whether or not equipartitioned.

It has long been observed that a matched beam with energy unbalance of up to around 50% can be injected "without much effect" on the emittances. Also, if a strongly energy-unbalanced (but matched) beam is injected, the distribution will at first move strongly toward equipartition. But as the EP ratio approaches some value in the vicinity of one, the EP ratio will abruptly level off and not proceed further toward one, or may oscillate around one. From the low-order mode charts of Hofmann discussed in [3], this behavior is not inconsistent with the low-order mode thresholds, and freedom from low-order instability in the EP region.

This behavior is quite well explained if one turns to the literature of complexity, nonlinear dynamics, and chaos. Accelerator beam dynamics is a typical nonlinear system, in which stochasticity thresholds appear as a function of the strength of a nonlinearity parameter. If the nonlinearity, e.g. lack of energy balance, is above a strong stochasticity threshold, very strong mixing (chaos, turbulence) occurs and the system is driven toward an equilibrium. Such mixing can be loosely described as a "thermalization" process, and effective settling times can be short.

When the above system equilibrates such that the nonlinearity strength falls below the strong stochasticity threshold, the dynamics changes fundamentally.

Now the transient settling time becomes very long, approaching infinity. Quasi-equilibria, or meta-equilibria, may still exist. In practical terms, for example, in the case of mismatched beams as discussed in [4], the conditions under which particles actually escape from the beam core onto the large mismatch orbits, the time intervals that particles spend in the core or in the halo, etc., are all random (while still a function of the nonlinearity strength). This property accounts for the observation that the exit time for particles to assume mismatch orbits tends to infinity for a mismatch approaching zero. Or, in the case of non-EP'd injection, the distribution approaches the EP condition until the threshold is crossed, whereupon the setting time becomes very long.

Most modern high-intensity linac (and ring) designs, whether equipartitioned or not, tend to lie in regions of tune-space where the nonlinearity strengths are smaller than the strong stochasticity threshold. (Older linacs of LAMPF vintage violate the threshold at low energy, and have strong emittance growth in this region.)

Below the strong stochasticity threshold, it is not proper to characterize the behavior as "thermalization". The beam would be in equilibrium if completely nonlinearly matched and energy balanced, but this equilibrium will not be a "thermal" one.

The characteristics of long settling times, then, makes it fundamentally quite unclear whether, when, how, etc., the EP condition may or may not be of benefit in accelerator design. (It should be pointed out that the time scales for intense, relatively short linacs and less intense but long ring channels are equivalent.)

Many details of this question are being explored outside the accelerator field by researchers in complexity, nonlinear dynamics and chaos, applied to classical mechanics, astrophysics and other areas. Basically, the question is whether the ergodic hypothesis is true below the strong stochasticity threshold.

RFQ TESTBED FOR STUDY OF EQUIPARTITIONING

Let us accept that an equilibrium design condition, even if only an rms one, is desirable, and discuss how systematic investigation might proceed.

The author's preference is to test the design in a simulation of a real accelerator, rather than constructing a simplified model. The RFQ has been chosen as the testbed. This is probably the most complex situation, because the beam is dc at injection and becomes bunched. Difficult as it may be, this is preferred because it forces attention to real factors. The RFQ is also a quite generic linac; for example, when a frequency jump is made between linac sections, the beam becomes long longitudinally and often the external fields would be shaped to shorten the bunch length again. On other ways, the RFQ is a simpler system, in that its fields are easy to describe analytically.

The RFQ generation is performed in steps. First, specifications are given at the end of a shaper (e.g., $\varepsilon_{ln}/\varepsilon_{tn} = 2$, $\phi_s = -88°$, aperture - $\beta\lambda/4$ along with the current, frequency and injection energy), and the modulation and beam dimensions are solved for to satisfy the matching and EP conditions. The injection section from input to the shaper end is then designed according to rules. The beam is to be equipartitioned from the shaper end through the rest of the RFQ; at each step (typically 1/10th of a cell), all parameters except the modulation and beam dimensions are found by solving the matching and EP equations simultaneously and exactly. (To date, machine layouts by others have approached the EP condition only approximately by trial and error, which compromises interpretation of the results.)

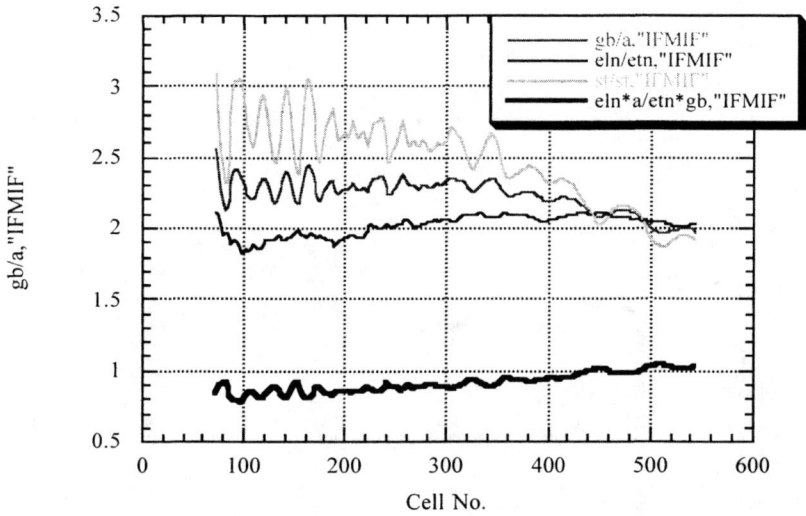

FIGURE 1. Result of RFQ design using Eqns. (1-3) with ellipsoidal form factor.

A typical result of this procedure is shown in Fig. 1 for a 140 mA, 100 keV-16 MeV deuteron RFQ (8 MeV final energy is considered practical in this case). The EP ratio is not well satisfied along the RFQ, although it tends to 1. The desired individual ratios $\varepsilon_{ln}/\varepsilon_{tn} = \gamma b/a = \sigma^t/\sigma^l = 2$ are also not well satisfied.

This is not too surprising, given the complex longitudinal dynamics. The possible contributions of several effects needed to be investigated. The matching equations (1) and (2) used for accelerator design are actually the equations for a time-invariant periodic transport system. Full envelope equations including the acceleration terms were derived; for RFQs whose parameter changes change slowly enough, as in most extant designs, these terms are small enough to be neglected. Higher-order terms such as would come from rf gaps or quad edges are not a problem in the RFQ.

As especially the longitudinal distribution is changing radically, it is clear that the geometry factor *ff* is changing in a complicated way, and the usual Sacherer approximation (ellipsoidal distribution, the same in r and z directions) is not valid. Additional information is needed, and is obtained from the first design run. A linear correction term for the form factor can be found from the ratio of (total longitudinal beam size/rms longitudinal beam size)/(total transverse beam size/rms transverse beam size). The numerator of this quantity varies from a more cylindrical beam distribution to a parabolic distribution; the denominator is always close to parabolic but does vary systematically.

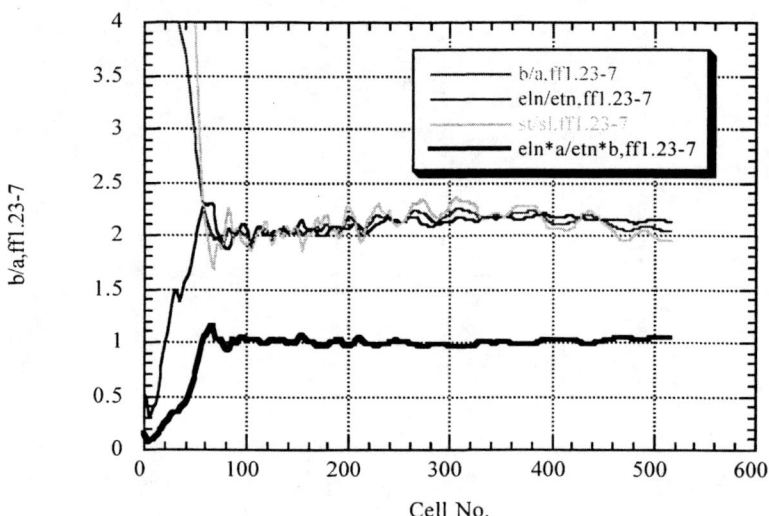

FIGURE 2. Result of RFQ design using Eqns. (1-3) with linear (with Cell No.) correction to the ellipsoidal form factor.

Application of this correction results in equipartitioning to about ±5%, with individual ratios to about ±10%, as shown in Fig. 2

Other schemes involving the kinetic, potential, space-charge, and nonlinear field energy terms also yield form factor corrections, typically requiring fitting to a 3rd-order polynomial, and also leaving a residual, systematic EP error of about ±5%.

Because the remaining deviation from the design requirement is systematic, and because the goal of the present work is to understand how to achieve the EP condition along the accelerating channel as closely as possible, it is desired to find a higher-order correction. The required form can be seen from numerical experiments, but has not yet been found analytically. The well-known phenomenon that a beam will seek a new distribution in a channel which self-consistently mirrors the external field is clearly happening, and expression of correction terms in this form, as explored by Batygin in this Workshop, is being pursued.

NATURE OF CHAOS

A final comment on the nature of chaos may be of interest. We have indicated herein and in [9] that resonances play a key role. Space-charge causes wide tune spreads (especially considering the local, instantaneous single-particle tune). Parametric resonances related to the focusing structure are expected to be the most powerful. An infinite number of other resonances exist at all rational number fractions; it is seen that these are used in particle transport. In circular machines, the whole set of re-entrant resonances is also introduced. Coherent effects cause shifts in the resonant frequencies as a function of the nonlinearity strength.

Until just a few years ago, the fundamental underpinnings of chaos theory were all based on perturbation theory, and it was known that the perturbation had to be extremely small for the theory to be valid. Yet the indicators, for example the Liapunov coefficients, worked quite well even for strong nonlinearity. This was troubling when reading the literature.

Techniques for geometrically modeling dynamical systems also have a long history. Chaos researchers looked at these and immediately understood that chaos could easily occur on surface manifolds with negative curvature. But they thought chaos could not occur if the curvature was positive. It was realized only a few years ago that this conclusion was wrong. Chaos can also occur if a positively curved manifold has "bumps" on it. And these bumps turn out to be the resonances! A revolution of sorts is now occurring in the literature, toward advanced geometrical methods for surface studies of general Hamiltonians, including time-varying, and deviations from the intended surface. The methods are rigorous for strong nonlinearities. It is interesting that resonances, so familiar in accelerators, are the main mechanism, and apparent that these new approaches should be of major relevance to accelerator designers.

REFERENCES

[1] OED/NEA Workshop on "Utilization and Reliability of High Power Proton Accelerators", October 13-15, 1998, Mito, Japan, Proc. to be published by JAERI.

[2] "IFMIF - International Fusion Materials Irradiation Facility Conceptual Design Activity, Final Report", M. Martone, editor, IFMIF-CDA Team (R.A. Jameson, Accelerator Facility Team Leader), ENEA Frascati Report, RT/ERG/FUS/96/11 (December 1996)..

[3] R.A. Jameson, "On Scaling & Optimization of High Intensity, Low-Beam-Loss RF Linacs for Neutron Source Drivers", AIP Conf. Proc. 279, ISBN 1-56396-191-1, DOE Conf-9206193 (1992) 969-998, Proc. Third Workshop on Advanced Accelerator Concepts, 14-20 June 1992, Port Jefferson, Long Island, NY, LA-UR-92-2474, Los Alamos National Laboratory.

[4] R.A. Jameson, "Self-Consistent Beam Halo Studies & Halo Diagnostic Development in a Continuous Linear Focusing Channel", LA-UR-94-3753, Los Alamos National Laboratory, 9 November 1994. AIP Proceedings of the 1994 Joint US-CERN-Japan International School on Frontiers of Accelerator Technology, Maui, Hawaii, USA, 3-9 November 1994, World Scientific, ISBN 981-02-2537-7, pp.530-560.

[5] J.S. O'Connell, et. al., Proc. 1993 Part. Accel. Conf., Washington, DC, May 1993, p. 3657.

[6] J.D. Meiss, Physica D74 (1994) 254-267.

[7] R. A. Jameson, "Beam-Intensity Limitations in Linear Accelerators," (Invited), Proc. 1981 Particle Accelerator Conf., Washington, DC, March 11-13, 1981, IEEE Trans. Nucl. Sci. 28, p. 2408, June 1981; R. A. Jameson, "Equipartitioning in Linear Accelerators", Proc. of the 1981 Linear Accelerator Conf., Santa Fe, NM, October 19-23, 1981, Los Alamos National Laboratory Report LA-9234-C, p. 125, February 1982; I. Hofmann, I. Bozsik, 1981 Linac Conf., LANL, LA-9234-C, pp. 116-119.

[8] I. Hofmann, "Equipartitioning and Halo Due to Anisotropy", Proc. 1997 Part. Accel. Conf.

[9] R.A. Jameson, "Beam Losses and Beam Halos in Accelerators for New Energy Sources", HIF Conf., Princeton, NJ, 6-10 Sept. 1995, Princeton, NJ, LA-UR-95-175, Los Alamos National Laboratory, September 1995. Fusion Engineering and Design 32-33 (1996) 149-157.

Numerical Code for Monte-Carlo Simulation of Ion Storage

Alekseev N., Bolshakov A., Mustafin E., Zenkevich P.

Institute for Theoretical and Experimental Physics, B.Cheremushkinskaya, 25, Moscow, Russia

Abstract. A numerical code for simulation of evolution of ion distribution functions during storage process has been proposed. The code based on Monte-Carlo method and «macroparticle» type simulation allows to estimate growth rates of the beam emittance and momentum spread due to an intrabeam scattering and an interaction with the stripping target for beams with non-Gaussian distribution function and permits to follow the process of particle losses as well. The algorithm of the code has been described, the code has been tested and results of its application to TWAC project have been given and discussed.

INTRODUCTION

Now in ITEP a new project of upgrade of the existing accelerator facilities called TWAC - **T**erra **W**att **AC**cumulator Refs.[1,2] is under construction. The first goal of the project is to store heavy ion beam with high accumulated energy in the storage ring based on the presently existing 10 GeV proton synchrotron U-10. The stored beam will be compressed, extracted and focused on the small size target to provide experiments on production of high density heavy ion plasma. The second goal is to create the accelerator complex suitable for studies and tests in heavy ion intense beams physics and technology.

In this project we assume to use a new scheme of non-Liouvillean multiturn injection. As it has been shown by preliminary analytical estimations Ref.[3] the main constraints to reach the necessary beam power (about 1 TW) arise from a growth of the longitudinal momentum spread due to intrabeam scattering which leads to energy transfer from the transverse plane to the longitudinal one and due to an interaction of the beam with the stripping target. Peculiarities of non-Liouvillean injection into the storage ring results in non-Gaussian distribution of the stored ions in a phase space. Long time of the ion circulation in the ring in absence of any cooling mechanism may lead to considerable ion losses during the storage process.

To investigate these effects we have developed computer code based on Monte-Carlo method and «macroparticle» type simulation. The injection scheme and the first results of numerical simulation of the accumulation process in TWAC by use of this code have been given in Ref.[4]. Here we describe in more details physical considerations and the structure of the code, as well as the last results of simulation of the beam accumulation process in TWAC.

NON-LIOUVILLEAN MULTITURN INJECTION

The principle of the injection is the following: injected beam of helium-like heavy ions (for example, $_{59}Co^{25+}$) penetrates through the stripping foil. The fully stripped ions of $_{59}Co^{27+}$ (about 80% of the primary beam) are captured into the stored beam by means of local bump of closed orbit of the storage ring specially created in order to the stripped ions fit the closed orbit. During the injection some ions in the stored beam could penetrate the foil and interact with its matter. Possible effects of this interaction include charge exchange due to electron pickup, multiple Coulomb scattering, ionization losses and inelastic collisions with the nuclei of the foil.

Disturbing effect of the foil on the beam is supposed to be radically reduced by minimizing total foil thickness penetrated by particles of the stored beam during the accumulation process. The technique to be used is the following: stripping target is placed outside of the stored beam which is moved into the beam only during the injection process (duration of this process is about one revolution period of ions in the storage ring). A kicker magnet system is installed in the storage ring to make a local bump of the closed orbit.

Additional minimization of a number of the ion interactions with the foil is supposed to be reached by taking into account the fact that the transverse phase volume of the stored beam is larger than injected beam emittance by two orders of magnitude, thus only small part of stored beam penetrates the foil during the injection The scheme of transverse phase space filling is shown on Fig.1. Looking through the Mendeleev Table we've taken fully striped ions of $_{59}Co^{27+}$ as a suitable specie from point of view of the beam dynamics and accumulator parameter definitions.

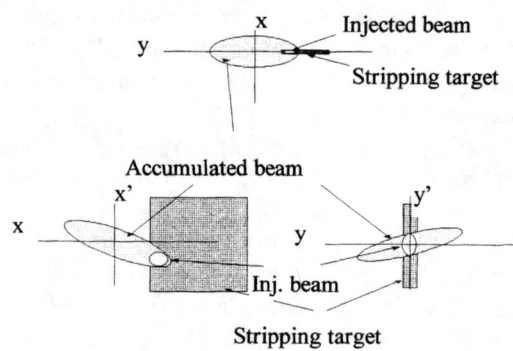

FIGURE 1. Transverse phase space filling during the multiturn injection into the storage ring.

Moreover Cobalt type ions are heavy enough for effective use in high density plasma production and are feasible for stripping to nearly bared state (+24 or +25 charge states) in a laser ion source as well. The main parameters of the beam to be injected into the storage ring are listed in Tab.1.

ALGORITHM OF THE CODE

The computer simulation is aimed to calculate the beam losses and to investigate an evolution of the particle distribution in 5D phase space (transverse degrees of freedom and momentum deviation) during accumulation process with account of initial distribution function of the injected beam, scheme of phase space filling at injection, beam interaction with stripping target and intrabeam scattering (IBS). Really, instead of this 5D phase space (which in the further text will be called «standard» phase space) we consider an evolution in the space of invariants namely the vector $I=\{I_1,I_2,I_3\}$, where I_1 is the longitudinal invariant, I_2, I_3 are the vertical and horizontal ones, correspondingly. The invariants are defined by

$$I_1 = \Delta p/p$$
$$I_2 = \gamma_x x_\beta^2 + 2\alpha_x x_\beta x_\beta' + \beta_x (x_\beta')^2$$
$$I_3 = \gamma_y y_\beta^2 + 2\alpha_y y_\beta y_\beta' + \beta_y (y_\beta')^2 \quad (1)$$

Here $\Delta p/p$ is the longitudinal momentum deviation, γ_x, α_x, β_x are the horizontal Twiss parameters, x_β and x_β' are given by

$$x_\beta = x - D(z)\frac{\Delta p}{p}, \qquad x_\beta' = x' - D'(z)\frac{\Delta p}{p} \quad (2)$$

In Eq.(2) D and D' are, correspondingly, the periodic dispersion function of the storage ring and its derivative. Equations for vertical oscillations one obtains from Eq.(2) replacing x by y and taking into account that $D=0$ and $D'=0$ for vertical motion.

Simulation algorithm is based on the method of macroparticles: the beam is considered as an ensemble of macroparticles defined in the space of invariants, and at each step of the calculations we consider a motion of each macroparticle. The process of beam evolution during the accumulation is treated as a sequence of the following

Table 1. Parameters of injected beam

Ion type	$_{59}Co^{25+} => _{59}Co^{27+}$
Kinetic energy, GeV	40
Intensity, ppc	10^{10}
β/γ	0.816 / 1.729
Momentum spread	±0.05%
Horizontal emittance, mm mrad	5 π
Vertical emittance, mm mrad	3 π
Storage ring hor. acceptance, mm mrad	110 π
Storage ring vert. acceptance, mm mrad	60 π

steps:
- Stored beam ensemble is supplemented by particles from injected beam. Truncated Gaussian distribution is used for definition of betatron amplitudes and momentum deviations of the injected macroparticles relatively to the center of the injected beam; the phases of the transverse oscillations are assumed to be distributed uniformly. Using this assumption we calculate the initial values of the vector I for the injected macroparticles.
- The betatron phases of the stored macroparticles are supposed to be distributed uniformly in interval of [0, 2π]; their values are chosen by use of uniformly random algorithm. Then their coordinates in the standard phase space are calculated
- Particles penetrating the stripping target are identified. New coordinates of these particles (and the particles of the injected beam) in standard phase space and corresponding new values of the invariants are been found with account of Coulomb scattering and energy loss in the target, as well as «effective» IBS collisions in this point. Particle losses due to electron pickup and nuclear interaction in the target are simulated by random exclusion of some macroparticles from the beam with an account of the known loss probability.
- Simulation of IBS in time intervals between injection cycles is applied to the stored beam by division of the ring superperiod on a sequence of points and successive simulation of the process in each point, where each particle «interacts» with another randomly chosen particle in correspondence with special algorithm which provides the same effect as «real» multiple intrabeam scattering.
- At each step of calculation particles with invariants exceeding some maximal value ($I_j > I_j^{max}$) are excluded from ensemble (being lost).

Let us consider in more details the interaction with the stripping foil.

Stripping: The yield of charge fractions of relativistic ions penetrating through the foil is determined by competition between electron stripping and pickup, and depends on projectile energy, foil matter and thickness. The yield of bare ions is increased with projectile energy and becomes independent of foil thickness after a sufficient (equilibrium) thickness is traversed. In order to minimize multiple Coulomb scattering the equilibrium thickness has to be used for stripping. The electron stripping cross section is nearly proportional to the second power of charge state of target nuclei (Z^2), and the specific weight of material is proportional to Z, thus the equilibrium thickness is approximately inversely proportional to Z.

Multiple Coulomb scattering: As it is well-known, due to multiple Mollier scattering derivatives x' and y' are changed randomly, and its changes $(\delta x')_{sc}$, $(\delta y')_{sc}$ obey a Gaussian probability law:

$$[(\delta x')_{sc}, (\delta y')_{sc}] = \exp\{[-(\delta x')_{sc}^2 - (\delta y')_{sc}^2]/2\sigma_{sc}^2\}/2\pi\sigma_{sc} \qquad (3)$$

with the dispersion

$$\sigma_{sc} = (15 \text{ MeV}/pc\beta)(\delta l/L_R)^{1/2} \qquad (4)$$

where pc is the particle longitudinal momentum in MeV/c, δl is the target length, L_R is the radiation length (constant of the material).

For each particles penetrating the target matter values of $(\delta x')_{sc}$, $(\delta y')_{sc}$ are randomly chosen in accordance with written above Gaussian law.

Ionization losses: The reduction of a particle longitudinal momentum due to ionization losses $\delta p/p$ is assumed to be constant (it is possible to show that influence of straggling is negligible). Then corresponding change of free horizontal oscillations due to the jump of the momentum is defined by

$$\delta x_\beta = -D(z)\,\delta p/p, \qquad \delta x'_\beta = -D'(z)\,\delta p/p \qquad (5)$$

We assume that the mean energy losses due to the ionization will be matched with the storage ring by slow diminution of the accumulator magnetic field. However, random character of the particle interaction with the target will result in increase of the beam momentum spread.

Complete changes of particle coordinates and its derivatives due to both effects are defined by:

$$\delta x_\beta = -D(z)\,\delta p/p, \qquad \delta x' = (\delta x')_{sc} - D'(z)\delta p/p,$$
$$\delta y = 0, \quad \delta y' = (\delta y')_{sc} \qquad (6)$$

Electron pickup and nuclear interaction: Estimations of the electron pick-up cross-sections have shown that the corresponding particle losses are negligible (see Table 1). Particle losses due to nuclear interaction are defined by:

$$\frac{\Delta N}{N} \approx A^{\frac{2}{3}} \frac{\Delta t}{L_N}$$

Here Δt is the target length (in g/cm^2), L_N is nuclear length of the target matter for protons, A is the atomic number of the ion. The numbers describing target effects are given in Table 2. These numbers (for three target materials) have shown that the disturbance of the stored beam by the foil is not extraordinary and doesn't make a serious obstacle for accumulation of thousand and more beam batches. The best material for recharge target has to be found in light elements like Aluminum or near it. If probability of ion to collide with the foil is about 0.04 (per injection cycle) and the mean life time of the ion in the ring chamber is equal to 500 cycles of injection, then the mean number of intersections with the target is about 20; for Al the corresponding losses due to electron capture and nuclear interaction are about 2%.

Now let us consider in more details the IBS simulation

Table 2. Foil influence on the penetrating beam

Target material	Mylar	Al	Cu
Equilibrium thickness for 80% bare ion yield, mg/cm^2	5	3	1.5
Ionization loss, MeV	8	3,9	1,8
Momentum loss, x10^{-2}%	1.2	0.6	0.2
Rms angle of multiple scattering, mrad 10^{-2}	7	7	7
Electron pickup cross section (non-radiative), barns	0.2	3.5	135
Electron pickup cross section (radiative), barns	1.0	1.8	4.1
Loss probability due to electron capture x10^{-4}	2.5	3.5	20
Loss probability due to nuclear interaction x10^{-4}	12.4	6.4	2.6

The straight forward simulation of IBS by macroparticles is impossible due to very large number of events ($10^7 \div 10^{13}$ interactions per second). To avoid this difficulty the «quasi-continuous» random process is replaced by an equivalent discrete random process. It is well known that the IBS process can be described correctly by use of Focker-Planck equation. From general point of view our discrete process is equivalent to IBS if the following conditions are satisfied:

1) change of invariants at each step of equivalent process $\Delta I_j << \Delta I_j^{max}$ (only in this case it is possible to use Focker-Planck equation for description of the discrete process);

2) the mean rates of changes of invariants ($\left\langle \dfrac{dI_j}{dt} \right\rangle$, and $\left\langle \dfrac{d(I_j I_k)}{dt} \right\rangle$, $j, k = 1,2,3$)

are the same for IBS and chosen discrete process.

Kinematics of the IBS is considered in details in Ref.[5]. According to Piwinski

$$\delta p/p = [\alpha \gamma \sin\Psi \sin\phi + \gamma \xi (\cos\Psi - 1)/2]$$

$$\frac{\delta p_y}{p} = \frac{1}{2}\{[\theta\sqrt{1+\frac{\xi^2}{4\alpha^2}}\sin\phi - \frac{\varsigma\xi}{2\alpha}\cos\phi]\sin\Psi + \varsigma(\cos\Psi - 1)\} \qquad (7)$$

Here $\gamma\xi = \delta p_1/p - \delta p_2/p$, $\theta = x_1' - x_2'$, $\varsigma = y_1' - y_2'$, $2\alpha \approx \sqrt{\theta^2 + \varsigma^2}$ is the angle between the incident particles in the laboratory frame. Ψ and ϕ are scattering and azimuthal angles in the center of momentum (CM) system between the particles after the collision Distribution on scattering angles ϕ and Ψ is given by the Rutherford cross-section:

$$d\bar{\sigma} = (\frac{r_i}{4\bar{\beta}^2 \sin^2\frac{\Psi}{2}})^2 \sin\Psi d\Psi d\phi \qquad (8)$$

where classical ion radius $r_i = (q^2/A)\, 1.53\cdot10^{-16}$ cm, $\bar{\beta} = (\beta\gamma/2)\,[\,(\Delta p_1/p - \Delta p_2/p)^2/\gamma^2 + (x_1'-x_2')^2 + (z_1'-z_2')^2\,]^{1/2}$

Eqs.(7,8) permit to calculate the mean rates of change for all invariants due to IBS. Let us consider, for simplicity, the mean rate of change for vertical invariant of «probe» particle with index «1». In accordance with Piwinski's calculations

$$\left\langle \frac{dI_3}{dt} \right\rangle = \left\langle \frac{\pi r_i^2}{4\gamma^2 \bar{\beta}^3} c\rho L_c(-4z_1'\varsigma + \xi^2 + \theta^2) \right\rangle \qquad (9)$$

Here ρ is the local density of the beam in the laboratory frame, L_c is the Coulomb logarithm ($L_c = \ln\dfrac{2\bar{d}\bar{\beta}^2}{r_i}$, $\bar{d} = \min(h, r_d)$, where h is the chamber half height, r_d is Debye radius).

We consider the discrete random process a sequence of «equivalent» scattering events with scattering angle Ψ_{eff} which is defined by:

$$\Psi_{eff} = \frac{2\sqrt{\pi}r_i}{(\bar{\beta})^{3/2}\gamma}\sqrt{c\cdot\rho\cdot\delta t\cdot\ln\frac{2d(\bar{\beta})^2}{r_i}} \qquad (10)$$

Here δt is a time interval between discrete «scattering events» in the equivalent process. We assume that kinematics of our «scattering event» is the same as for real intrabeam scattering act, and scattering angle ϕ is distributed randomly by an uniform probability law within the interval of $[0,2\pi]$. Substitution of this value of Ψ_{eff} into Eq.(9) and calculation of $\left\langle \dfrac{dI_3}{dt} \right\rangle$ has shown that values of mean rates for both processes are the same. Corresponding calculations confirm that this equivalence takes place (with an accuracy of L^{-1}, where Coulomb logarithm L is about 20) for all last derivatives.

It is necessary to mark that if $\bar{\beta} \to 0$ then $\Delta I_j \to \infty$. Such events are very rare, if δt is chosen small enough; however, to avoid such events we have imposed the following limitation on Ψ_{eff}: $\Psi_{eff} < \Psi_0$, where Ψ_0 is a free parameter of the code.

Now let us describe the formal features of the IBS simulation algorithm.

1) The superperiod of magnetic system is treated as a sequence of N equidistant points; Twiss parameters, dispersion function and its derivative are calculated for each point. Simulation procedure is executed successively for each point of the sequence through interval δt, which is considered as a free parameter of the code.

2) For each point the betatron phases of the macroparticles are distributed uniformly by use of random algorithm, then the macroparticle coordinates in standard phase space are calculated. Aperture of the vacuum chamber is divided by grid on separate cells, and for each particle the it is identified a cell which contains it; then the local density inside all the cells is calculated.

3) For each particle it is chosen a «partner» located in the same cell. «Effective scattering angle» Ψ_{eff} for given «partners» is calculated by use of the procedure described above; the second scattering angle ϕ is chosen randomly within the interval of $[0,2\pi]$. New values of the invariants for these two particles are calculated, and then they are excluded from the further analysis in this step of the code. If one of these invariants exceeds the corresponding maximal value, the macroparticle is considered to be lost.

The chosen free parameters of the code are the following ones: 1) number of injected macroparticles (typically about 1000); 2) number of cells per aperture of the vacuum chamber (about 100); 3) number of the injection cycles (up to 1000); 4) maximal effective scattering angle Ψ_0 (about 0.05). It has been found a weak dependence of the simulation results on variations of these parameters.

Validation of the code has been made by comparison with results obtained by the code of Giannini and Moehl Ref.[6], the last one permits to find change of the invariants for Gaussian distributions only. Results given in Table 3 (for number of ions $N_{ion}=10^{12}$) show a satisfactory coincidence.

Table 3. Comparison with the code of Gianinni and Moehl

Time, sec	σ_x	σ_z	σ_p^2	Growth time, sec	
				The code	G&M
0	$2.27 \cdot 10^{-5}$	$1.43 \cdot 10^{-5}$	$4.75 \cdot 10^{-8}$		
0.2	$2.27 \cdot 10^{-5}$	$1.43 \cdot 10^{-5}$	$5.57 \cdot 10^{-8}$	2.51	2.33
0.5	$2.27 \cdot 10^{-5}$	$1.44 \cdot 10^{-5}$	$6.78 \cdot 10^{-8}$	2.81	2.78

RESULTS

Firstly, processes of the stored beam interaction with the stripping foil and IBS effect have been simulated separately to evaluate their comparative contribution into the beam dynamics. Main results are presented in Fig.2-4 where evolution of the transverse emittance and beam momentum in the accumulating ring is seen. Calculations were done for the aluminum foil of 3 mg/cm^2 thickness and one second interval between injections. The influence of the foil penetration is shown for 1000 injection cycles, the IBS considered for 300 sec. The loss factor at the end of simulation is 8% for the foil interaction at 1000 injections and about 10% for the IBS at 300 cycles of injection. It's seen very clear (Fig.4) sufficiently strong effect of momentum dispersion growth for the IBS. The accumulated beam parameters arising from the IBS effect after 300 cycles of injection are the following: accumulated intensity is about $3 \cdot 10^{12}$, horizontal emittance for 90% of beam is less than 80 π mm mrad, vertical emittance is less than 30 π mm mrad, momentum spread is about ±0.25%. The last value is the most crucial for accumulation because it limits the final beam compression.

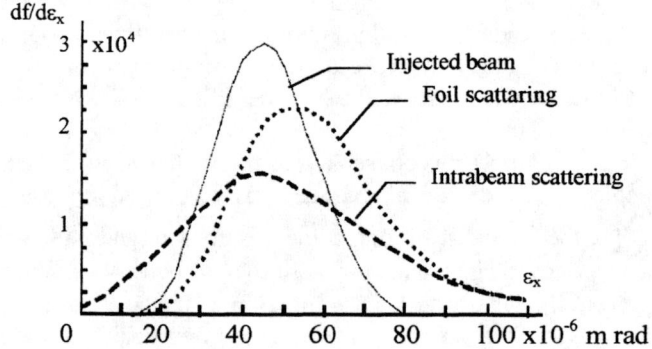

FIGURE 2. Horizontal emittance growth due to beam scattering in foil and IBS ($A_x = 1.1\ 10^{-4}$ m rad).

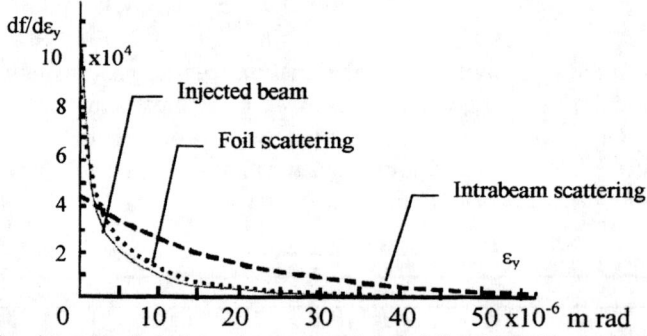

FIGURE 3. Vertical emittance growth due to beam scattering in foil and IBS ($A_y = 0.6\ 10^{-4}$ m rad).

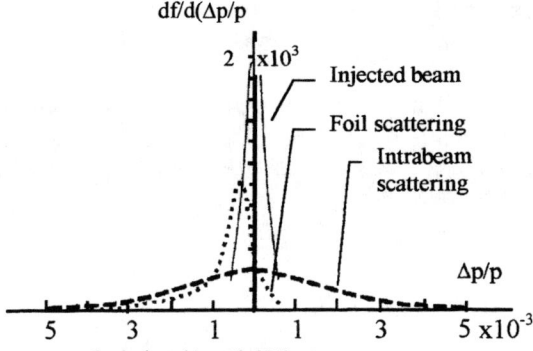

FIGURE 4. Distribution of momentum deviation ($A_z = \pm 0.8\%$)

At last, we have made calculations of the beam evolution with account of both effects. At Fig.5 the evolution of the particle distribution in the space of horizontal invariant is plotted for 200 (2.4% particles are lost), 400 (19%) and 600 (39%) cycles of injection (number of injected ions is 10^{10} per cycle). Corresponding evolution of the momentum spread is plotted in Fig.6 (for smoothed distributions).

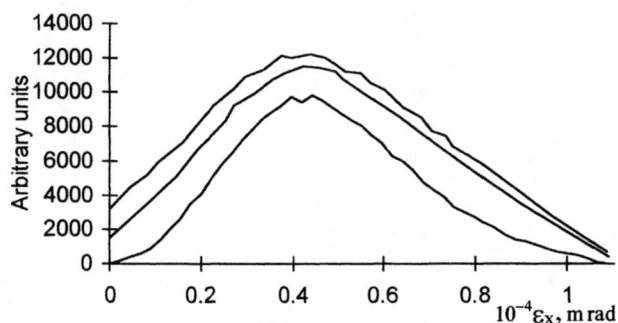

FIGURE 5. Evolution of ion distribution in the space of horizontal invariant (200, 400, 600 sec)

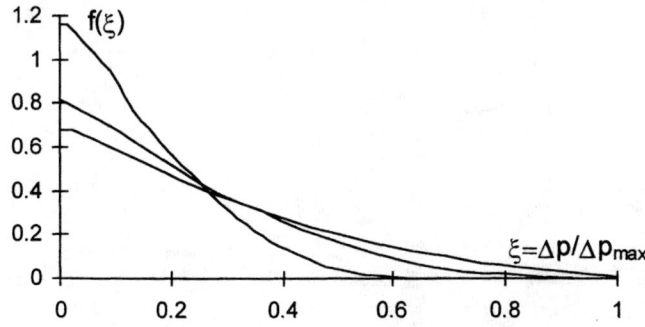

FIGURE 6. Evolution of ion distribution in a space of momentum deviation.

We see that longitudinal momentum spread is continuously increased due to transfer of the energy from «hot» transverse degrees of freedom into the longitudinal one.

Known distribution functions permit to calculate longitudinal and transverse stability diagrams plotted, correspondingly, in Figs.7,8. We see that beam stability is firstly improved due to increase of the beam momentum spread and appearance of «tails» in distribution function. However, when losses of ions in longitudinal phase space appears (that means that $f(1)\neq 0$) the longitudinal coherent oscillations become unstable.

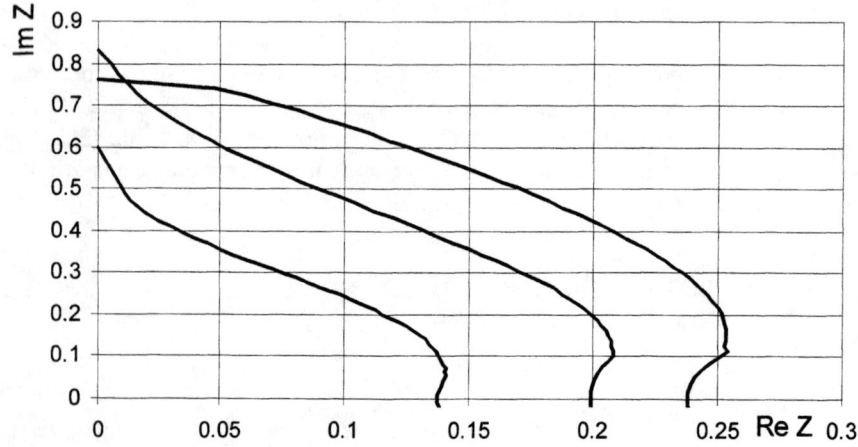

FIGURE 7. Diagram of longitudinal stability.

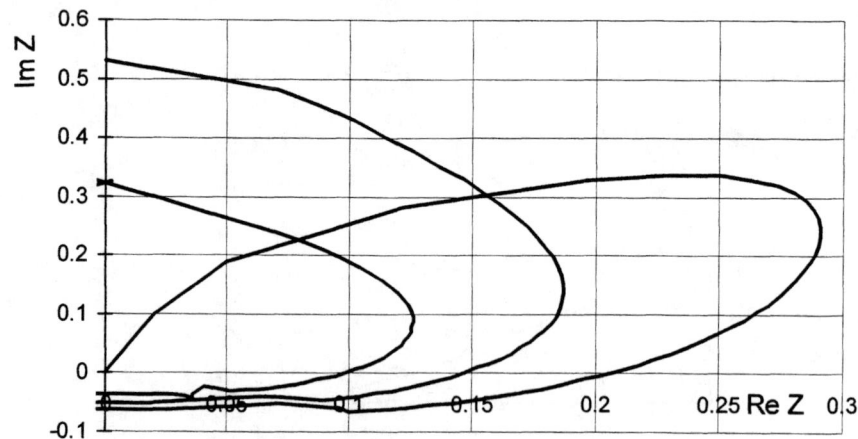

FIGURE 8. Diagram of transverse stability

CONCLUSION

Computer code based on macroparticle method has been elaborated for simulation of ion beam accumulation with account of IBS and interaction with the stripping foil. The method of IBS simulation is based on replace of quasi-continuous IBS process to equivalent discrete random process, which may be described by Focker-Planck equation with the same diffusion rates (and other coefficients) as the ones for the initial IBS process. This discrete process represents a sequence of effective scattering events with similar kinematics as initial process of intrabeam scattering; scattering angles are chosen by such a way to provide an equivalence of both processes. Validation of the code by comparison with the code of Giannini and Moehl Ref.[6] has confirmed its satisfactory accuracy.

The first results of simulation performed for the scheme of accumulation of Cobalt ions in the ITEP-TWAC ring give estimation of utmost value for intensity, emittances and momentum spread of stored beam. Excessive growth of the beam momentum spread due to IBS suppresses the beam coherent instabilities, but results in significant particle losses and makes difficult high compression of stored beam in the time domain. Forthcoming work is to consider possibilities to improve the accumulation efficiency by optimization of the injection process, increase of the repetition rate and implementation of electron cooling.

In conclusion we would like to mark that Monte-Carlo method proposed here could be applied, with some modifications, for simulation of similar processes, such as electron and stochastic cooling. Moreover, we believe that proposed algorithm can be included in PIC codes.

REFERENCES

1. D.G. Koshkarev et all., *Proc. of XV Russian Workshop on Charged Accelerators, Protvino,* **2**, p.319, 1996.
2. B.J. Sharkov et all, *NIM A,* **415**, N 1, pp.20-26, 1998.
3. E.R.Mustafin, P.R.Zenkevich, «Collective Effects in Designed ITEP Heavy Ion Complex», presented at the Sarantsev's Memorial Symposium, Dubna, 1997 (forthcoming).
4. N.N.Alexeev et al, *Proc. of EPAC'98, Stokholm,* p.1147, 1998.
5. A.Piwinsky, *Proc. of 9th Int. Conf. on High Energy Accelerators,* p.105, 1974.
6. R.Giannini, D.Moehl, *INTRAB, a Computer Code to Calculate Growth Rates, Emittance Evolution and Equilibrium with Intrabeam Scattering and Cooling,* PS/AR note 92-22,1993.

The injection and storage schemes for heavy ion beams

I. Meshkov[1], E. Syresin[1], T. Katayama[2] and Y. Yano[3]

[1]*Joint Institute for Nuclear Research, Dubna, Moscow region 141890, Russia*
[2]*Center for Nuclear Study, Graduate School of Science, University of Tokyo, 3-2-1, Midorich, Tanashi, Tokyo 188, Japan*
[3] *The Institute of Physical and Chemical Research (RIKEN), Hirosawa 2-1, Wako, Saitama 351-01, Japan*

Abstract . Two injection schemes: multi-turn injection with RF stacking and a single turn injection are discussed in this report. The multi-turn injection scheme has an advantage in the number of turns, however, it causes to cooling of a large beam emittance before stacking[1]. On the other hand, the single turn scheme has a significant gain in the ion storage rate, if the initial emittance of the injected beam is small[2,3]. We begin with the general consideration of both schemes and then give numerical estimations demonstrating characteristic features of these schemes for ACR. The problem of the ion source choice is also discussed.

INJECTION OF ION BEAMS INTO THE STORAGE RING

Few schemes are used for injection of the ion beams into storage rings. A multi-turn injection with RF stacking has an advantage in numbers of turns. The disadvantage of this scheme is a large emittace of injected beam. A stripping injection is often used for injection of the ion beams into storage rings. The restrictions for application of this scheme are related to the limited number of ion types used for this scheme and also to the ion life time. A single turn injection with an electron or a stochastic cooling has a significant gain in the ion storage rate for the injected beams with a small initial emittance. Realization of this scheme is essentially determined by availability of the high intensive pulse source of highly charged ions. The single turn injection with cooling and RF stacking can be realized also in a multi-bunch mode.

SINGLE TURN INJECTION SCHEME

The scheme alternative to multi-turn injection[1] is the single turn injection with storing of the cooled particles in a RF separatrix, that produces a bunch of stored particles[2,3]. In this injection scheme the electron cooling method can be realized with high efficiency due to a high quality of the injected beam. The loss of intensity caused by reduction of turns may be compensated by a more intensive ion pulse source. There are several kinds of the intensive pulse sources of highly charged ions, so called an electron-beam ion source EBIS in the reflection mode[5-7] and a laser ion source[8-9]. The advantage of this scheme is fast cooling of the injected ion beam with small initial emittance. This scheme presumes simultaneous cooling of an injected beam and its bunching as soon as the beam becomes cold.

The momentum spread of the injected ions is higher than the value of the momentum spread at which the ions can be captured in the RF separatrix (Figure 1). During electron cooling the ion momentum spread becomes small, and ions became captured in the RF separatrix. At this moment a new portion of ions can be injected in the free space of the ring. And definitely, the kicker pulse duration is to be shorter of the revolution period τ_{rev}, and the pulse itself is to be synchronized with the RF station in order to avoid a perturbation of the stored beam.

The RF frequency in the storage ring has to correspond to the first harmonic of the operation mode. The single turn injection can be realized at a high RF harmonic of the operation mode. There is a gap between ion trajectories of the injected ion beam and the stack and free space in the ring, that is required for the new ion portion, is appeared after RF stacking.

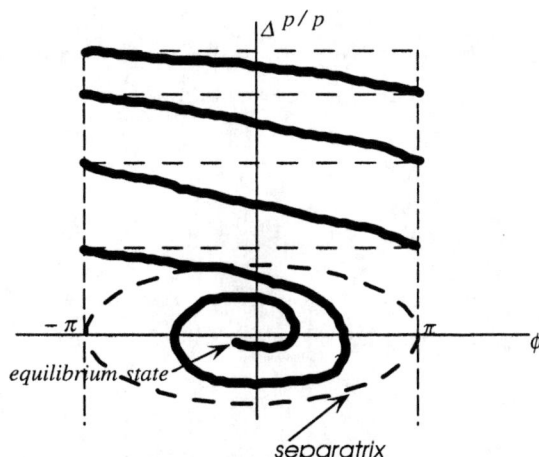

FIGURE 1. The ion dynamics in the longitudinal phase space under the influence of electron cooling and RF voltage.

Below we consider the single turn injection on the first harmonic. The part of the ring circumference occupied by the stored beam is defined by the beam momentum spread $\Delta p_{cool}/p$ and RF voltage U_{RF}:

$$\Delta \phi_{stored} = \sqrt{2\pi\eta_\omega h \beta^2 \gamma \frac{A}{Z} \cdot \frac{m_p c^2}{eU_{RF}} \cdot 2 \cdot \frac{\Delta p_{cool}}{p}}, \quad \eta_\omega = \frac{1}{\gamma^2} - \frac{1}{\gamma_{tr}^2}, \quad (1)$$

h is the harmonic number, β, γ are the relativistic factors, A and Z are the atomic mass number and the charge number of the ion, m_p is the proton mass, c is the speed of light, U_{RF} is RF voltage, γ_{tr} is γ-transition.

The more detailed scenario that seems acceptable for the single turn injection is described as the following. The RF voltage rapidly grows to its maximum value. Following a quarter of synchrotron phase oscillation, when the bunch has the minimum length, new portions of ions is injected. Then the RF voltage returns to the initial level at the moment when the beam momentum spread has the minimum value. The ion momentum spread of the injected beam is reduced during the cooling time, after which an injection of a new portion of ions is possible. By performing repetitions of the single turn injection cycles it is possible to accumulate the cooled beam in the stationary RF bucket.

The number of ions stored in the ring during one pulse of the single turn injection is equal to

$$N_{st}^{Single} = \dot{N}_{Single} \tau_{kick}, \qquad (2)$$

where N_{Single} is the ion beam intensity, τ_{kick} is the time, when kicker is switched on. This time has to be compared with the revolution time τ_{rev} in the ring, and we believe, that $\tau_{kick} \approx \tau_{ref}$. Intensity of the ion beam corresponds to

$$\dot{N}_{Single} = \frac{n_b^{Single}}{\tau_{rf}}, \qquad (3)$$

where n_b^{Single} is the number of ions per bunch at a single turn injection, $\tau_{rf} = 1/f_{rf}$, f_{rf} is the accelerating RF frequency. Now one can define the storing rate R as the following:

$$R_{Single} = \frac{1}{\dot{N}_{Single}} \cdot \frac{N_{stored}^{Single}}{\tau_{cool}^{sin gl}} = \frac{\tau_{rev}}{\tau_{cool}^{Single}} \qquad (4)$$

The electron cooling time $\tau_{cool}^{sin gle}$ in the ACR storage ring corresponds to 30-200 msec. This value is comparable with the time of RF stacking and stochastic cooling at the multi-turn injection. For the single turn injection, both electron and stochastic cooling systems can be used effectively. For both systems, the cooling times are comparable if numbers of ions in the ring are close to 10^7. Both cooling systems can be operated simultaneously in order to obtain additional gain in the cooling time.

The total number of stored ions N_{tot}^{Single} after the period of life time τ_{life} is determined by the balance of the supply rate and the decay rate and it is equal to

$$N_{tot}^{Single} = n_b^{Single} \frac{\tau_{rev}}{\tau_{rf}} \frac{(1-e^{-1})}{(\exp(\tau_{cool}^{Single}/\tau_{life})-1)} \qquad (5)$$

The number of the stored ions in the ring is given by

$$N_{tot}^{Single} = n_b^{Single} \frac{\tau_{rev}}{\tau_{rf}} \frac{\tau_{lifel}}{\tau_{cool}^{Single}} \qquad (6)$$

at $\tau_{cool}^{Single}/\tau_{life} \ll 1$.

COMPARISON OF THE SINGLE TURN AND MULTI-TURN INJECTION SCHEMES

The use of the RF stacking technique[1] performs the accumulation of the unstable beam in the storage ring. The momentum cooling works continuously during the RF stacking. The cooling time depends on the quality of the ion beam. The emittance of ion beam at the multi-turn injection is large. The electron cooling cannot be so effectively as the stochastic cooling at a weak ion intensity.

The total number of the RI ions stored in the ring after the period of intrinsic life time τ_{life} at the multi-turn injection is equal to:

$$N_{tot}^{multi} = n_b^{multi} N_{inj} \frac{h}{D} \frac{(1-e^{-1})}{(\exp((\tau_{cool}^{multi}+\tau_{st})/\tau_{life})-1)}, \qquad (7)$$

where n_b^{multi} is the number of ions per bunch at the multi-turn injection, N_{inj} is the number of turns, D is the dilution factor, τ_{st} is the stacking time, τ_{cool}^{multi} is the cooling time at the multi-turn injection, h is the number of RF harmonic at multi-turn injection, $h = \tau_{rev}/\tau_{rf}$, τ_{rev} is the revolution time, $\tau_{rf} = 1/f_{rf}$, f_{rf} is the accelerating RF frequency.

The number of stored ions in ACR at

$$(\tau_{cool}^{multi}+\tau_{st})/\tau_{life} \ll 1 \qquad (8)$$

is given by

$$N_{tot}^{multi} = n_b^{multi} N_{inj} \frac{\tau_{rev}}{\tau_{rf} D} \frac{\tau_{life}}{\tau_{cool}^{multi}+\tau_{st}}, \qquad (9)$$

The ratio between the ion beam intensity produced at the single turn and at the multi-turn injection schemes corresponds to

$$K = \frac{N_{tot}^{Single}}{N_{tot}^{multi}} = \frac{n_b^{Single}}{n_b^{multi}} \frac{1}{N_{inj}} \frac{\tau_{cool}^{multi} + \tau_{st}}{\tau_{cool}^{Single}}. \qquad (10)$$

At the first view the multi-turn injection is preferable to the single turn injection due to a gain in numbers of turns N_{inj}. But the effective use of both electron and stochastic cooling systems at a small initial emittance of the single turn injected beam gives a gain in cooling time in comparison with the multi-turn injection, when the ion beam has a large emittance. The second, even more essential gain is the possibility to choose one pulsed ion sources[5-9] with a higher intensity than the intensity of a ion source operating in the continuous mode[4].

ELECTRON COOLING IN THE MULTI-TURN AND SINGLE TURN INJECTION SCHEMES

In this section we describe the analytical simulations of the storage rate ratio in the multi-turn and single turn injection schemes that use the electron cooling method. We start this comparison with consideration of the multi-turn injection scheme. A beam of intensity $\dot{N}_{multi}$ and of emittance ε_i is injected into a cooler-storage ring. If the multi-turn injection scheme with N_{inj} turns is applied, the current and the emittance of the circulating beam will be N_{inj} times larger, than of the injected one, and the total number of particles in the circulating beam is equal to

$$N_{st}^{multi} = N_{inj} \dot{N}_{multi} \tau_{rev}, \qquad (11)$$

where τ_{rev} is the particle revolution period in the ring.

It is well known that the electron cooling time is proportional to the cube of the particle angular spread Θ_i or to the beam emittance in the power 3/2, when $\Theta_i \gg \Theta_e$, where Θ_e is the angular spread of the cooling electron beam. Such situation is typical for a storing process. Therefore, one can write:

$$\tau_{cool} = k\varepsilon_i^{3/2}, \qquad (12)$$

where k is a constant, which has the dimension $[k] = s \cdot cm^{-3/2}$. For the multi-turn injection this gives

$$\tau_{cool}^{multi} = kN_{inj}^{3/2} \varepsilon_i^{3/2}. \qquad (13)$$

Therefore, substituting (11) and (13) into (4) we obtain the storing rate:

$$R^{multi} = N_{inj} \frac{\tau_{rev}}{\tau_{cool}^{multi}} = \frac{\tau_{rev}}{k\sqrt{N_{inj}}(\varepsilon_i)^{3/2}}. \quad (14)$$

The single turn injection enables to essentially reduce the cooling time, but also causes a loss of the intensity of the ion beam in the storage ring of factor N_{inj} comparing with the multi-turn injection. In this case, the circulating (to be cooled) beam emittance is equal to that of the injected beam. Therefore, the cooling time at the single turn injection is equal to

$$\tau_{cool}^{Single} = k\varepsilon_i^{3/2} \quad (15)$$

and the storage rate, correspondingly, is

$$R^{Single} = N_{inj} \frac{\tau_{rev}}{\tau_{cool}^{Single}} = \frac{\tau_{rev}}{k(\varepsilon_i)^{3/2}}. \quad (16)$$

Considering the (15) and (16) we conclude, that the single turn injection scheme gives the gain in the storing rate of factor of $\sqrt{N_{inj}}$, comparing to the multi-turn scheme

$$R^{Single}/R^{multi} = \sqrt{N_{inj}}. \quad (17)$$

The pulse ion source with a high intensity gives an additional gain at the single turn injection in the combination with electron cooling method:

$$\frac{R^{Single}}{R^{multi}} = \sqrt{N_{inj}} \frac{\dot{N}_{pulse}}{\dot{N}_{cw}}, \quad (18)$$

where $\dot{N}_{pulse}$ and $\dot{N}_{cw}$ are the intensity of a ion source in the pulsed and continuous modes, respectively.

NUMERICAL SIMULATION OF THE ELECTRON COOLING TIME AT THE SINGLE TURN INJECTION

The qualitative estimations presented above leave some questions of the reliability of the proposed single turn injection scheme. Main question is how strongly RF voltage influences the electron cooling process, because it initiates ion phase oscillations, which cause, in principle, an increase of ion beam momentum spread. The numerical simulations are presented for the ACR storage ring[1].

To answer this question, the numerical simulation of the injection regimes was produced using the "BETACOOL" program[10] on the base of the analytical formulaes[11]. It allows to compute ion motion in a cooler-ring taking into account an influence of the ring characteristics like focusing structure and betatron function magnitudes in the cooling section, the longitudinal magnetic field there and the RF voltage (the bunched beam regime).

The parameters chosen for computation (Table 1) correspond to the examples[1] as an illustration of the multi-turn injection scheme. The computation shows that if RF voltage is less than 1 kV, the cooling time practically does not increase with the voltage amplitude. The simulation of longitudinal cooling time is given in Figure 2b and Figure 3b. The damping of the synchrotron oscillations during cooling, when RF voltage of $U_{RF}=1$ kV is applied, is shown in these figures.

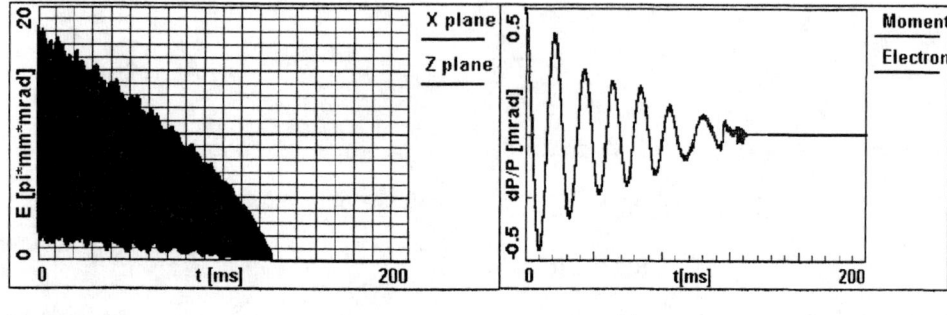

a) b)

FIGURE 2. The numerical simulations of the electron cooling by BETACOOL program at single turn injection of ions $^{132}Sn^{50}$ in ACR; a) the beam emittance, d) the momentum spread. The ion beam parameters: E_i=210 MeV/u, $\varepsilon_h / \varepsilon_v = 10/10$ $\pi \cdot mm \cdot mrad$, $\Delta p_{inj} / p = 0.5 \cdot 10^{-3}$. The electron beam parameters: electron beam current I=4A, electron beam diameter d=5cm.

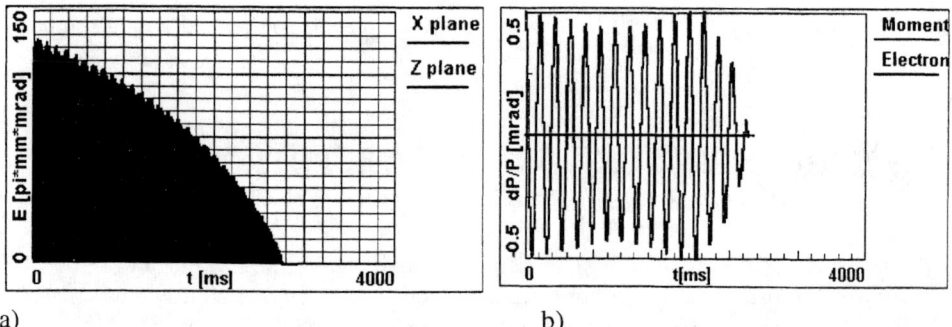

a) b)

FIGURE 3. Numerical simulations of the electron cooling by BETACOOL program at the multi-turn injection of ions $^{132}Sn^{50}$ in ACR a) the beam emittance, b) the momentum spread. The ion beam parameters: E_i=210 MeV/u, $\varepsilon_h / \varepsilon_v = 125/10$ $\pi \cdot mm \cdot mrad$, $\Delta p_{inj} / p = 0.5 \cdot 10^{-3}$. The electron beam parameters: I=4A, d=5cm .

TABLE 1. The cooler ring and beams parameters

ACR parameters	
Circumference, m	168.4836
$B\rho$, T m	7.244
Q_x / Q_y	4.555 / 3.54
γ_{tr}	4.987
l_{cooler} (cooling section length), m	3.6
$\eta = l_{cooler} / C$	0.021
(β_x / β_y) in cooling section, m	5.97 / 12.24
(D_x / D_y) in cooling section, m	0 / 0
RF harmonic	1.0
RF voltage, kV	0.5 – 15
Parameters of the injected ion beam	
Ion	$^{132}Sn^{50+}$
Ion energy, MeV/u	210
Emittance $\varepsilon_x = \varepsilon_y$, π mm mrad	10
Initial momentum spread $\Delta p_{inj} / p$	$(0.5\text{-}1.5)*10^{-3}$
Ion beam size and angular spread in the cooling section	
(x_0 / y_0), cm	0.77 / 1.1
(Θ_x / Θ_y), Mrad	1.3 / 0.9
Number of turns at multi-turn injection	12
Emittance after multi-turn injection $\varepsilon_x / \varepsilon_y$, π mm.mrad	125 / 10
Ion beam size and angular spread in the cooling section after multi-turn injection	
(x_0 / y_0), cm	2.73 / 1.1
(Θ_x / Θ_y), mrad	4.57 / 0.9
Electron beam parameters	
Current, A	4.0
Energy, keV	217.8
Electron beam radius, cm	2.5
Electron transverse temperature, meV	40
Electron beam angular spread, mrad	0.275

The cooling time of the full emittance of the single turn injected ion beam $^{132}Sn^{50}$ is about

$$\tau_{cool}^{Single} \approx 125 \text{ msec,}$$

just as the cooling time of full beam emittance at the multi-turn injection is 21 times longer:

$$\tau_{cool}^{multi} \approx 2650 \text{ msec.}$$

These numbers give us the storing rates (see (4) and (17)):

$$\frac{R^{Single}}{R^{multi}} = \frac{1}{N_{inj}} \frac{\tau_{cool}^{multi}}{\tau_{cool}^{Single}} \approx 1.75,$$

where $N_{inj}=12$ is the number of turns. Numerical simulations are in a practical agreement with the analytical estimation (17) $R^{Single}/R^{multi} = \sqrt{N_{inj}} = 3.4$. This simulations are performed at initial momentum spread of $\Delta p_{inj}/p \approx 0.5 \cdot 10^{-3}$. The gain in storing rate, obtained from numerical simulations, is comparable with one at a higher momentum spread of $\Delta p_{inj}/p \approx 1.5 \cdot 10^{-3}$.

The storage rate gain resulting from a decrease of the electron beam radius

The application of the single turn injection in ACR at a low ion beam emittance allows also to decrease the electron beam radius and to increase correspondingly the electron current density. The injected ion beam with emittance $\varepsilon_h/\varepsilon_v = 10/10 \; \pi \cdot mm \cdot mrad$ has transverse size less than 2.2 cm. The use of the variable radius electron beam, as foreseen in the project[1], allows to tune the electron beam size to an optimal value. It is realized by the variation of the magnetic field in the gun superconducting solenoid with maximal value of B=5T. For $^{132}Sn^{50}$ the cooling time is 2 times less (Figure 4) comparing with the date given in the Figure 2 for the electron beam with diameter of 5 cm. The storage rate gain at the single turn injection is

$$\frac{R^{single}}{R^{multi}} = 3.3,$$

when the number of turns is $N_{inj}=12$.

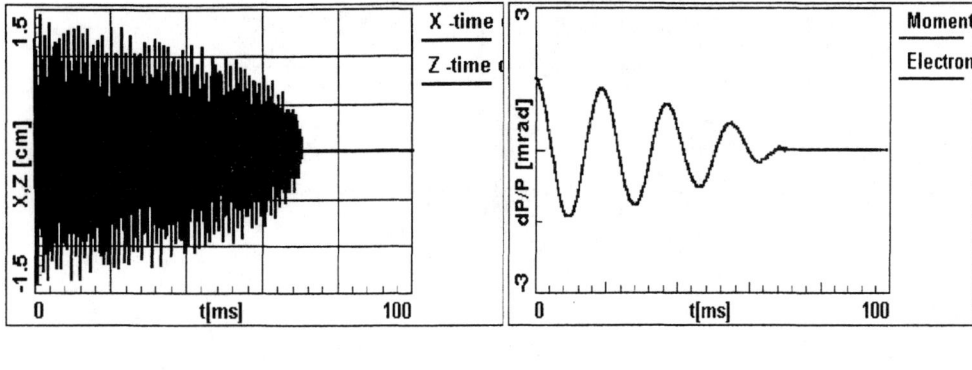

FIGURE 4 a) The transverse cooling of ions $^{132}Sn^{50}$ b) The longitudinal cooling. The ion beam parameters: E_i=210 MeV/u, $\varepsilon_h/\varepsilon_v = 10/10 \ \pi \cdot mm \cdot mrad$, $\Delta p_{inj}/p = 1.5 \cdot 10^{-3}$. The electron beam parameters: I=4A, d=2.5cm

The cooling of ions by the dense neutralized electron beams at single turn injection

The cooling time is reduced 2 times if the cooling is done by the neutralized electron beam (Figure 5). The gain in the cooling time for the neutralized electron beams[12-13] is realized at the small initial momentum spread of the ion beam $\Delta p_{inj}/p \approx (0.5-1) \cdot 10^{-3}$. The neutralization of the electron beam space-charge is very important for the fast cooling of injected ion beams with good quality: small initial momentum spread and beam emittance. The cooling time by the neutralized electron beam corresponds to 30 ms. It is compared with stacking time and cooling time by the stochastic method. The storage rate gain at single turn injection and the electron cooling with neutralized electron beam of small diameter is equal to

$$\frac{R^{single}}{R^{multi}} = 6.8,$$

when number of turns is N_{inj}=12.

The generation of the dense neutralized electron beam is restricted by a beam-drift instability[13]. The threshold electron beam current of the development beam-drift instability is higher than 5 A for the parameters given in Table 1 and beam diameter of 2.5 cm.

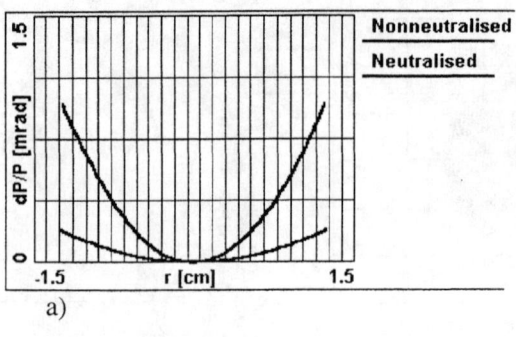
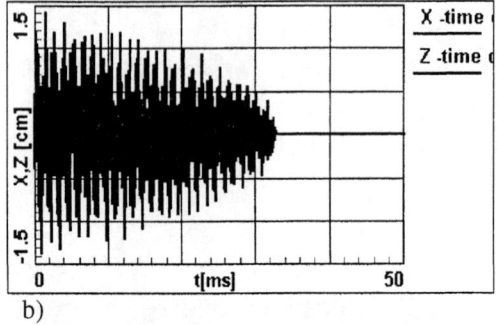

FIGURE 5. a)The radial potential distribution inside the charged and 80% neutralized electron beams. b)The cooling of $^{132}Sn^{50}$ by the 80% neutralized electron beam. The ion beam parameters: E_i=210 MeV/u, $\varepsilon_h/\varepsilon_v = 10/10\ \pi \cdot mm \cdot mrad$, $\Delta p_{inj}/p = 0.5 \cdot 10^{-3}$. The electron beam parameters: I=4A, d=2.5cm

COMPARISON OF THE ELECTRON AND STOCHASTIC COOLING METODS AT THE SINGLE TURN INJECTION SCHEME IN ACR

The stochastic cooling is very effective in comparison with the electron cooling in the multi-turn injection due to a larger horizontal ion beam emittance and rather weak ion intensity[1] (Figure 6).

Here we compare both cooling methods at the single turn injection, when the injected ion beam has the small emittance of $\varepsilon_h/\varepsilon_v = 10/10\ \pi \cdot mm \cdot mrad$ and initial momentum spread of $\Delta p_{inj}/p \approx 10^{-3}$. Both cooling methods give low cooling time for number of particles less than 10^8, compared with RF staking time at the multi-turn injection (Figure 6). The neutralization of the electron beam gives a higher gain at case of use of the electron cooling for the low energy ions (Figure 7). The single turn injection with use of both cooling methods provides a possibility to reach the injection repetition frequency in ACR about 20 –50 Hz.

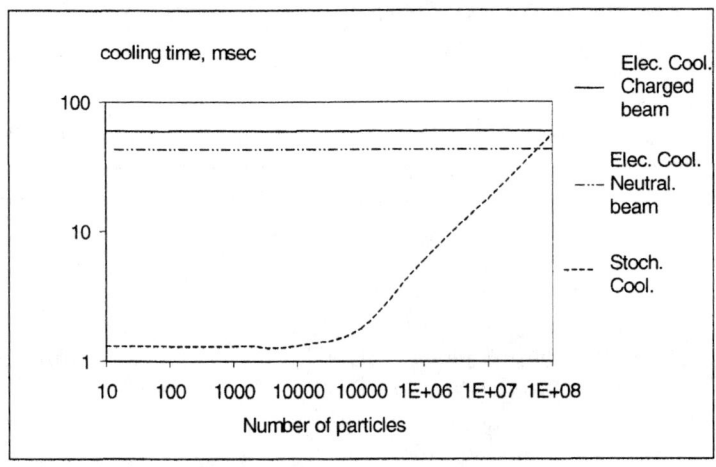

FIGURE 6 Dependence of the cooling time in the ACR on the number of ^{32}Sn^{50}ions. The ion beam parameters: E_I= 210 MeV/u, $\varepsilon_h/\varepsilon_v = 10/10$ π mm mrad, $\Delta p_{inj}/p = 10^{-3}$. The electron beam parameters: I=4A, d=2.5 cm, η=0.8 for the neutralized beam, η=0 for the charged beam. The stochastic cooling parameters: P_{max} =10 kW is the microwave power, w=2GHz is the band width.

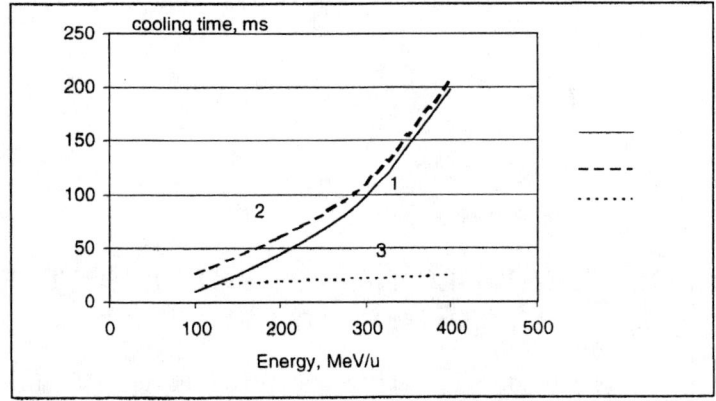

FIGURE 7. Dependence of the cooling time on the ion energy for ^{32}Sn^{50} beam. $\varepsilon_h/\varepsilon_v = 10/10$ π mm mrad $\Delta p_{inj}/p = 10^{-3}$. Curve 1 is the electron cooling by the neutralized

beam with $\eta=0.8$, curve 2 is the electron cooling by the charged beam, curve 3 is the stochastic cooling, $N_i=10^7$.

The electron cooling time at the single turn injection in ACR corresponds to the stochastic cooling time at number of ions of 10^7

$$\tau_{cool}^{electron} \approx \tau_{cool}^{stoch} \approx 30 \text{ ms.}$$

As a result, the application of both cooling systems at the single turn injection gives the 2 times gain in the cooling time comparing with the multi-turn injection and the stochastic cooling. The cooling and stacking times are compared at the multi-turn injection:

$$\tau_{stack} \approx \tau_{cool}^{stoch} \approx 30 \text{ ms.}$$

The stacking is done simultaneously with the cooling in the single turn injection, it gives an additional 2 times gain for this scheme. The gain in the storage rate ratio for the ACR single turn injection plus both cooling methods (electron and stochastic) compared with the multi-turn injection plus stochastic cooling corresponds to

$$\frac{R_{electron\&stoch}^{single}}{R_{stoch}^{multi}} = \frac{\dot{N}_{pulse}}{\dot{N}_{CW}} \frac{1}{N_{inj}} \frac{\tau_{single}}{\tau_{multi}} \approx \frac{\dot{N}_{pulse}}{\dot{N}_{CW}} \frac{1}{N_{inj}} \times 4.$$

The typical ratio between the pulsed and continuos intensities in a ion sources can reach the level of $\dot{N}_{pulse}/\dot{N}_{cw} \approx 10$. The gain in the storage rate ratio in this case is estimated as

$$\frac{R_{electron\&stoch}^{single}}{R_{stoch}^{multi}} \approx 2-3$$

at number of turns $N_{inj}=10-20$.

SCHEME OF THE SINGLE TURN INJECTION ON THE HIGH RF HARMONIC

The scheme of single turn injection can be realized on the high RF harmonic. There is a gap between ion trajectories of the injected beam and the stack at the momentum shift between both beams. The free space required for the newly injected ion bunches appears after RF stacking. The ions are cooled after injection into the storage ring, then they come to the stack orbit during RF stacking. Two cooling systems are preferable for the simultaneous staking and fast cooling, when the momentum spread of the injected beam is essentially less than the momentum shift between the stack and injected beam

$$\Delta p_{inj} \ll \delta p_{stack}.$$

One cooling system operates at the top of the stack energy, other works at the energy of injected beam. As an example, the electron cooling works at the energy of injected beam for fast cooling. The stochastic cooling operates at the energy, that corresponds to the top of the ion stack. Also for the ACR the initial momentum spread of the injected beam is essentially less ($\Delta p_{inj}/p=0.0015$) than the momentum shift $\Delta p_{stak}/p \approx 0.02$ between the stack and injected beam. The stochastic cooling can be applied for stack at the ion energy that corresponds to the ion stack top. One can reach the fast cooling of the injected beam and effective stacking using two cooling systems. When one cooling system is realized at the energy of the ion stack top, the cooling time of the injected bunches is essentially increased due to a large momentum shift between stack and injected bunches.

The RF frequency f_{min} corresponds to a high harmonic h of the revolution frequency of the injected bunches. Then, after the cooling time passes, the RF frequency increases up to frequency f_{max} related to a high harmonic of stack revolution frequency. The RF stacking is realized at this operation. Then, the frequency returns back to the initial value. The RF frequency should be changed fast to prevent the return of a new portion of stacking ions to the injection orbit.

Modification of the electron cooling system for the single turn injection on the high RF harmonic

The single turn injection scheme on the high RF harmonic can be realized with a high efficiency by use of the electron cooling system without the stochastic cooling. For this purpose a few modification of electron cooling system are proposed in[3]:
1. The electron cooling system with two drift sections at different potentials and in zero dispersion section of the ring;
2. The electron beam with a gradient of longitudinal velocity;
3. The electron cooling system with two electron beams of different energy.

The application of an electron cooling system with two drifts sections (Figure 8) is preferable way for the simultaneous fast cooling of the injected and stacked beams. The injected and stacked ions in the electron cooling section have the same trajectories caused by zero dispersion in the electron cooling system. The cooling section is divided in two parts. The potential of vacuum chamber of part 1 corresponds to energy of injected beam, the potential of the vacuum chamber of part 2 is related to the top energy of stack ions, it is constant during cooling. The potential of the part 1 is changed during cooling of injected ions (Figure 9).

$$p_{stack} - p_{inj} \approx 10^{-2} \cdot p_{inj},$$

therefore, the stack passes through the part 1 of the vacuum chamber without an additional perturbation. The same regime realized for injected beam. It very effectively cooled in part 1 and does not obtain perturbation, when passes through the part 2.

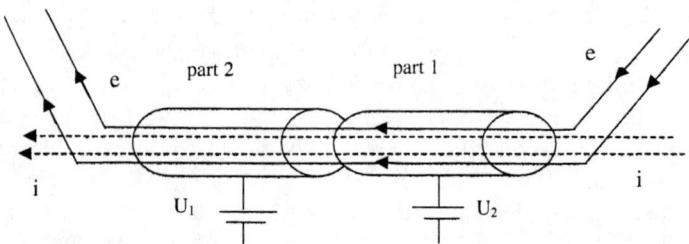

FIGURE 8. The scheme of the electron cooling system with two drift sections at different potentials and in zero dispersion section of the ring.

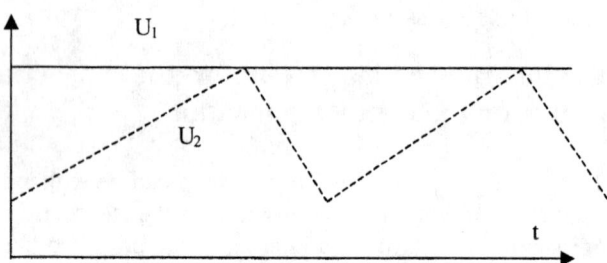

FIGURE 9. Dependence of the applied voltage to the both sections of the electron cooling system on time.
When an ion stack passes trough the part 2 of vacuum chamber it is effectively cooled. The momentum shift between the stack and injected beam is large

The other modification of electron cooling system is related to the generation of electron beam with the gradient of the longitudinal velocity. In this scheme the gap between ion trajectories of the injected beam and stack appears due to a dispersion in electron cooling section. The electron beam in electron cooling system is produced in a band or ellipse shape with the longitudinal velocity gradient (Figure. 10).

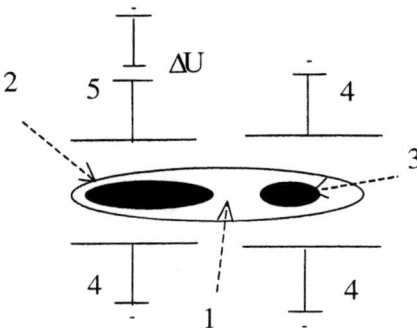

FIGURE 10. The simultaneous cooling of the ion stack and injected bunches by the electron beam with a gradient of longitudinal velocity. 1 is the electron beam, 2 is the ion stack, 3 is the new injected bunch, 4 is the vacuum chamber at zero potential, 5 is the additional electrode that produces the gradient of longitudinal velocity across the electron beam.

The injected beam passes through one edge of the electron beam, and stacking ions move through its opposite edge. The potential related to the momentum shift of stacking ions applied to electrode 5 (Figure 10). As a result, the electrons passed between electrodes 4 and 5 have a shift in the momentum comparing with the electrons moved between electrodes on the opposite edge of the beam. The electron beam with the gradient of longitudinal velocity provides vastly cooled ion stack and the new injected bunches simultaneously. The gain in the cooling time for the injected bunches in the scheme with the longitudinal gradient is realized due to small initial momentum spread of the injected bunches.

There is also another solution of the simultaneous fast cooling of the injected ion bunches and the ion stack in the electron cooling system. Two electron beams with diameter related to emittance of the injected beam with different momentum are produced in the electron cooling system (Figure 11).

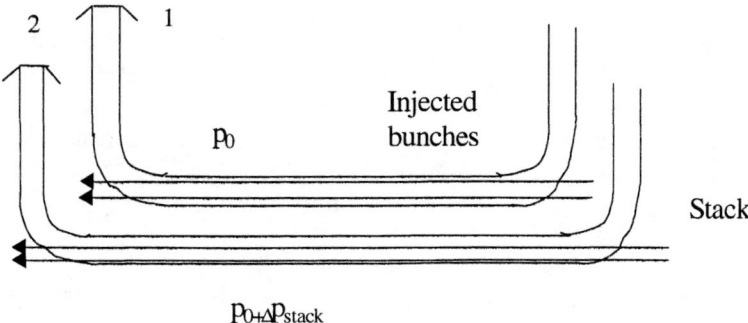

FIGURE 11. The simultaneous cooling of ion stack and injected bunches by two electron beams, produced in the electron cooling system.

The momentum of the electrons in one beam corresponds to the momentum of the injected ion bunches at the same time, in the other beam the electron momentum is equal to the stack top momentum. Two different cathodes can produce these two electron beams. The potential of one cathode differs from the potential of the other by the value, related to the momentum shift of stack respectively to the injected ion bunches.

The examples considered here show that several technical possibilities can be used to realization of the single turn injection on high RF harmonic with the fast simultaneous cooling of the ion stack and injected bunches in one electron cooling system.

FAST RESONANT STACKING

The adiabatic stacking requires to satisfy the condition[1]

$$\Omega \tau_{st} \gg 1, \tag{19}$$

where Ω is the synchrotron frequency. The stacking time τ_{st} determines the storage rate for multi-turn injection in (10). This time can be essentially reduced at a quasilinear stacking, if the ion bunch is cooled before the stacking. The RF frequency varies linearly in time (Figure 12).

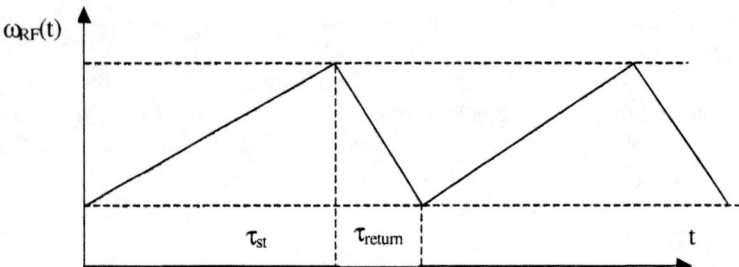

FIGURE 12. Dependence of the RF frequency on time for the quasilinear stacking.

It means, that the ion momentum p_s corresponding to ω_{RF} is changed in time in accordance with formula

$$x_s(t) = x_{max} \frac{t}{\tau_{st}}, \quad x_s = \frac{p_s(t) - p_s(t_{inj})}{P_{stack}}. \tag{20}$$

The synchrotron motion at the quasilinear stacking is described by the following equation:

$$\ddot{x} + \Omega^2 x = \Omega^2 x_s + \ddot{x}_s, \qquad (21)$$

where $x = (p_i - p_s)/p_{stack}$, p_i is the ion momentum. The dependence of ion momentum p_i on time during quasilinear stacking can be found from the equation (21) (Figure 13)

$$p_i = p_{inj} + (p_{stack} - p_{inj}) \cdot \left(\frac{t}{\tau_{st}} - \frac{\sin \Omega t}{\Omega \tau_{st}} \right). \qquad (22)$$

The fast resonant stacking is realized the resonance condition is regarded, when

$$\Omega \tau_{st} = n\pi, \qquad (23)$$

where n is integer. The momentum spread excited by the fast return of the RF frequency can be estimated as

$$\left(\frac{\delta p}{p} \right)_{return} \approx \frac{\Delta p_{stack}}{p} (\Omega \tau_{return})^2. \qquad (24)$$

The application of the quasilinear fast resonant stacking to the ACR ring reduces the stacking time down to several msec.

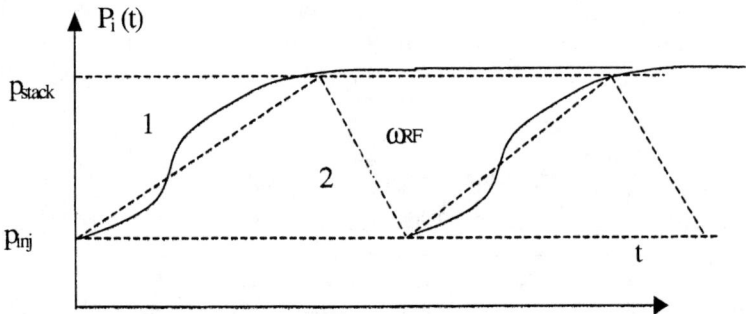

FIGURE 13. Dependence of the ion momentum (curve 1) and the RF frequency (curve 2) on time during quasilinear stacking.

THE CHOICE OF THE ION SOURCE

The advantage of the single turn injection scheme in the ion intensity gain appears when a high intense pulsed ion sources is used instead of one operating in continues mode. Thus the number of the ions stored in the ACR ring increases at single turn injection scheme if the ion intensity $\dot{N}_{pulse}$ of the pulsed ion source is higher by one order
of magnitude than intensity of the source operated in the continues mode $\dot{N}_{cw}$:

$$\dot{N}_{pulse} / \dot{N}_{cw} \approx 10.$$

So, the high intense ion beam in ACR the ca be obtained with multi-turn injection when ECRIS[4,14] is used. However this source generates ions with mass ratio of A/Z≈2.5-14.5. For further acceleration the proposed scheme requires A/Z≈2-3 and correspondingly the application of the Charge State Multiplier (CSM) plus an Stripper lines[14-15].

This scheme can be simplified when an intense pulsed ion sources such as EBIS in reflection mode[5-7] either laser source[8-9] is used with single turn injection regime.

The production of the highly charged heavy ions in the EBIS

The recent progress achieved in so called Electron-Beam Ion Sources (EBIS) originally developed in JINR[16] and modified later (so named EBIS in reflection mode[5-7]), makes possible an efficient application of these ion sources to single turn injection in the ACR. The EBIS in the reflection mode produces the highly charged heavy ions with A/Z≈2-3. The intensity of this source is essentially higher than one of the ECRIS, which is supposed for the generation of heavy ions of the mass-charge ratio of A/Z≈5. The peak intensity of Ar^{16+} beam achieved with EBIS[5-7] is of 1.2 eµA (Table 2).

TABLE 2. The comparison of the EBIS and ECRIS parameters

Device	18GHz ECRIS CW mode[4]	18 GHz ECRIS Pulsed mode[4]	EBIS[5-7]	EBIS[17] (Upgrate)
		Ar^{+8}		
Ion current, eµA	470	500		
		Ar^{11+}		
Ion current, eµA	160	180		
		Ar^{16+}		
Ion current, eµA	5	5	1200	5000
		Pulse parameters		
Pulse duration, µs		$(1-3)*10^3$	8	5
Repetition frequency, Hz		10	5	25

The number of Ar^{16+} ions from the EBIS in reflection mode is equal to $3.7*10^9$ ions/pulse at the pulse duration of 8 μs and repetition frequency of 5 Hz. It means that now the efficiency of Ar^{16+} generation in the EBIS is a few times higher, than production of Ar^{8+} in ECRIS (taking into account the striping up to Ar^{16+} with the efficiency of 40%):

$$\frac{\dot{N}_{EBIS}}{\dot{N}_{ECRIS+CSM}} \approx 7.5$$

The intensity of ions Ar^{16+} in the EBIS is 240 times higher comparatively to the intensity of Ar^{16+} obtained with the 18 GHz ECRIS.

The EBIS parameters are expected to be improved soon: the repetition frequency up to 25 Hz and the intensity - up to 5 times higher[17].

Laser ion source

The metallic ions are produced in 18 GHz ECRIS by MIVOC method[4]. The intensity of ions Fe^{10+} - Fe^{13+} is equal to 200 eμA for pulsed ECRIS mode[4]. It is 2 times less for continues mode[4]. The number of these ions produced by the ECRIS during the multi-turn injection time (of 30 μs) corresponds to $3*10^9$ ions. The intensity of the $^{73}Ta^{15+}$ and $^{73}Ta^{25+}$ corresponds to 50 eμA and 20 eμA, respectively. The metallic ions of a higher charge can be obtained at use of the ECRIS plus the CSM.

The pulsed production of metallic ion beam with very high intensity at relatively low charge state is realized in the laser sources[8-9]. The experimental data for JINR ion laser source are given in Figure 14 (Table 3). It means that now the efficiency of Ar^{16+} generation in the EBIS is a few times higher, than production of Ar^{8+} in ECRIS (taking into account the striping up to Ar^{16+} with the efficiency of 40%):

$$\frac{\dot{N}_{EBIS}}{\dot{N}_{ECRIS+CSM}} \approx 7.5 .$$

The number of metallic ions produced by the laser source per pulse is few times higher, than number of ions produced by ECRIS during the multi-turn injection time:

$$\frac{\dot{N}_{Laser}}{N_{inj} \cdot \dot{N}_{ECRIS}} \approx 30 .$$

The laser source[8] has a low repetition rate f_{Laser}, therefore its gain in intensity reduces by $(\tau_{cool}^{multi} + \tau_{stack})f_{Laser}$ times. As result of the laser source application, the number of ions stored in the ACR increases by several times for the single turn injection comparing with the multi-turn injection scheme:

$$\frac{N_{tot}^{sin\,gl}}{N_{tot}^{multi}} \approx \frac{\dot{N}_{Laser}}{N_{inj} \cdot N_{ECRIS}} \left(\tau_{cool}^{multi} + \tau_{stack}\right) f_{Laser} \approx 3-5.$$

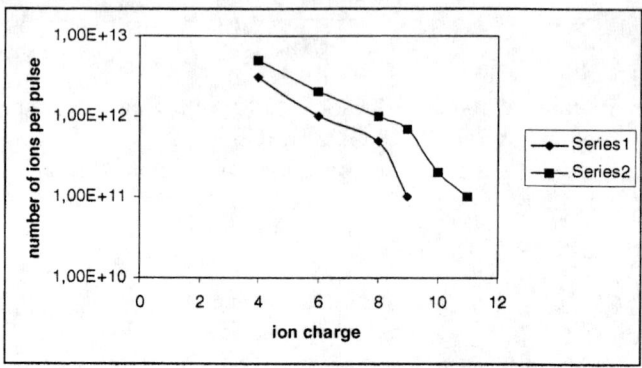

FIGURE 14. Dependence of the number of metallic ions per pulse on the ion charge [8-9]. The CO_2 laser source parameters: energy per pulse of 4 J, pulse width of 100 ns and 5- kV extraction pulse with repetition frequency of several Hz. Series 1 is for ^{23}V ions, Series 2 is for ^{73}Ta ions.

TABLE 3. The comparison of the laser ion sources[8-9] and 18 GHZ ECRIS[4]

Ion source	18 GHz ECRIS, CW mode[4]	18 GHz ECRIS, pulse mode[4]	Laser source[8]	Laser source[9]
		Ta^{15}		
Ion current, eµA	20	50		
Number of ions per N_{inj} turns injection	$1.7*10^9$	$4.3*10^9$		
Number of ions per turn	$8.6*10^7$	$2*10^7$		
		Ta^{11}		
Number of ions per turn			10^{11}	
		C^5		
Ion current, eµA	700	700		
Number of ions per N_{inj} turns injection	$6*10^{10}$	$6*10^{10}$		
Number of ions per turn	$3*10^9$	$3*10^9$	$2*10^{12}$	
		Al^5		
Number of ions per turn				$2*10^{11}$
Pulse duration, µs		$(1-3)*10^3$	0.3-0.5	
Repetition frequency, Hz		10	1	25

The highly charged metallic ions produced by the laser source can be obtained in the ACR with the CSM. The other way to obtain highly charged ions is the use of a powerful laser source of 100 J energy per pulse. For example, the laser source[18] permits to inject $9 \cdot 10^9$ Pb^{25+} ions/pulse at pulse duration of 6 µs.

CONCLUSION

The application of the single turn injection scheme together with a pulsed ion source (the EBIS, the laser source) gives the gain in the number of stored ions in the ACR ring, if the ion source intensity of the $\dot{N}_{pulse}$ is by one order higher than that one in continuous mode $\dot{N}_{cw}$ which is used for multi-turn injection:

$$\frac{\dot{N}_{pulse}}{\dot{N}_{cw}} \approx 10$$

The realization of the single turn injection together with high intense pulsed ion sources gives the gain in stored ions in the ACR by factor of 3-5 for typical cases, considered in this report.

The proposed scheme is especially important in the case of storing of short-lived isotopes.

REFERENCES

1. *MUSES Conceptual Design Report, RIKEN*, Chap.2, 1997
2. Parkhomchuk V. V., Ter-Akopian G. M., Meshkov I. N., Syresin E. M. at al., in "Heavy ion storage ring complex K4-K10. Technical Proposal", *JINR E 9-92-15, Dubna* (1992), pp.16-28.
3. Meshkov I., Syresin E., Katayama T. and Yano Y. "Single Turn Injection Scheme for ACR", *RIKEN-AF-AC-8*, 1998
4. Nakagawa T., Arje J., Miyazawa Y., Hemmi M. att al., "Recent Progress of RIKEN 18 GHz ECRIS", *Collected papers on the RIKEN RI beam factory project*. September 1997, p.99-104, 1997
5. Donets E. D "Review of recent developments for electron-beam ion sources (EBIS)", *Rev. Scien. Instrum.* 67(3), 873, (1996)
6. Donets E. D. "25 Years with EBIS", *Phys. Scripta*, 71, 5, (1977)
7. Donets E. D. *Rev. Scien. Instrum. Proc. of Intern. Conf. on ion* sources, Italy, Taormina, 1997
8. Anan'in O.B., Bykovskii Yu. A., Gusev V. P., Kolesov I.V. at al. *Sov. Phys. Tech. Pys.* 27, 903, (1982*), The physics and Technology of Ion sources*. Ed. G. Brown, Jon Wiley, 1989, p. 304
9. Bykovskii Yu. A.,. Gusev V. P., Kolesov I.V., Kutner V. B., *JINR, Dubna, Report P9-86-2* (in Russian) *The physics and Technology of Ion sources*. Ed. G. Brown, Jon Wiley, 1989, p. 306
10. Lavrentev A.Yu. and.Meshkov I. N., "The computation of electron cooling process in a storage ring", *Preprint JINR E9-96-347, JINR, Dubna* (1996).
11 Meshkov I. N., Phys. Part. Nucl.25 (6), 631, (1994).
12. Bosser J., Ley R., Meshkov I., Syresin E. et al., "Electron cooling with neutralised electron beams", *Proc. EPAC. London*, 1994, vol.2, p.1211

13. Bosser J., Korotaev Y., Meshkov I., Syresin E et al., "The active methods of instability suppression in a neutralized electron beam", *NIM A* 391 (1), 110, (1997)
14. Yano Y., Goto A., Miyazawa Y., Hemmi M. at al., "RIKEN RI beam factory project" *Collected papers on the RIKEN RI beam factory project and RIKEN ring cyclotro*, June 1996, p. 1-9, 1996
15. Kamigaito O. "Initial Design of Charge State Multiplier for RIKEN RI beam Factory", October 1997
16. Donets E. D. "Electron beam Ion sources", *The physics and Technology of Ion sources. Ed. G. Brown, Jon Wiley,* 1989, pp. 245-279
17. Donets E. D., Private communication
18. Sharkov B., "TWAC facility and the use of the laser ion source for production of intense heavy ion beams", *This proceeding*

Electron Cooling with Circulating Electron Beam in GeV Energy Range

I. Meshkov

Joint Institute for Nuclear Research (JINR), 141980,Dubna, Russia

Abstract. The method of electron cooling with an use of electron beam circulating in a storage ring is described. The thermodynamics of the cooling process is considered and scheme of the cooler ring with longitudinal magnetic field is discussed. The proposed method can be used for electron cooling of ions and antiprotons in GeV energy range.

1. INTRODUCTION

An extension of the electron cooling method into the GeV energy range is extremely desirable, because this allows to reach high luminosity in hadron and hadron-electron colliders to utilise used antiprotons and prepare high quality beams to be injected into collider (Recycler, PETRA); to maintain luminosity of the colliding beams (ENC, COSY) (Table 1), [1-4]. The most advanced one among them is Fermilab project [1] based on electrostatic "Pelletron" accelerator: the DC electron beam of 0.5 A current is obtained at electron energy of 1 MeV. The peculiarity of this device is the use of the same vacuum tube for acceleration and deceleration of electrons: two beams travel in opposite directions focused by permanent magnet lenses. That differs the Fermilab project with pioneering experiment at Budker INP in Novosibirsk, where two tubes device with magnetized electron beam was constructed in begining of the 80-th and DC electron beam of 1A current was obtained [5].

Nevertheless, an application of traditional scheme of electron cooler to cooling of heavy particles of e few GeV energy meets obvious problems of generation of high intense electron beams at energy of several MeV. One can avoid these problems using an electron beam circulating in a storage ring [1-4]. This method of cooling technique has own limitations considered in the presented report, particularly – the "thermocapacity" of cooling electron beam.

The principle scheme of the cooler (Fig.1) includes an injector ("electron gun"), storage ring and electron collector ("electron dump"). The ring has straight section inserted in the structure of the hadron storage ring, where electron beam merges with hadron one and cools it.

The electron ring can be used simultaneously for preliminary acceleration of electrons. Then the injector delivers electrons of rather low energy. So called "modified betatron" is one of possible schemes of the storage ring with cooling electron beam. Its application to the electron cooling is considered below. Combination of the ring with an injector provides an added bonus.

CP480, *Space Charge Dominated Beam Physics for Heavy Ion Fusion*,
edited by Yuri K. Batygin
© 1999 The American Institute of Physics 1-56396-860-6/99/$15.00

2. PECULIARITIES OF THE ELECTRON COOLING PROCESS WITH CIRCULATING ELECTRON BEAM

2.1. "Thermodynamics" of the cooling process in approximation of uniform two component system

Due to interaction between the particles (antiprotons, ions) and electrons the particle temperature decreases when the electron one increases. The variation of both temperatures in *Maxwellian plasma* is described by the equations:

$$\frac{dT_p}{dt} = \frac{4\sqrt{2\pi}\eta n_e z^2 e^4 L_C}{\gamma^2 mM} \cdot \frac{T_p - T_e}{\left(\frac{T_p}{M} + \frac{T_e}{m}\right)^{3/2}};\qquad(1)$$

$$N_p \frac{dT_e}{dt} = -N_e \frac{dT_e}{dt},\qquad(2)$$

where T_p, T_e are the particle and electron temperatures in the particle rest frame, m and M – their masses, N_p, N_e – the particle numbers in the rings, η is the ratio of the cooling section length to the circumference of the particle ring, L_C – Coulomb logarithm, n_e – the electron density, t – current time in Laboratory reference frame. The particle temperature in a cooler with *single pass electron beam* (and with Maxwellian velocity distribution!) decreases in accordance with the 1st equation, where $T_e = const$.

For typical parameters of the cooler projects (Table 1) the number circulating electrons N_e is comparable with that one of the particles N_p circulating in hadron collider.

TABLE 1. Cooler projects in GeV energy range

Parameter	Recycler FNAL	PETRA (DESY)	ENC (GSI/INP)	COSY (Juelich FZ)
Particles	antiprotons	Protons	ions	Protons
Energy, GeV (GeV/amu)	8	20	30	2.5
Ring circumference, m	3319.4	2304	1000	184
Particle number	$1.4 \cdot 10^{13}$	$3.8 \cdot 10^{12}$	$2 \times 10^{10}(p) \div 4 \times 10^7$ (U^{92+})	$3 \cdot 10^{10}$
Emittance, π·mm·mrad	3.3	4	$6.4 \times 10^{-3} \div 1 \times 10^{-3}$	2.5
Momentum spread, 10^{-3}	0.9	0.5		0.4
Electron beam				
Kinetic energy, MeV	4.36	10.9	16.3	1.3
Electron current, A	0.2 – 2	0.3 - 1	$8 \div 0.02$	0.2 - 2
Electron beam radius, cm	1.0	1.0	$0.11 \div 0.04$	3.2

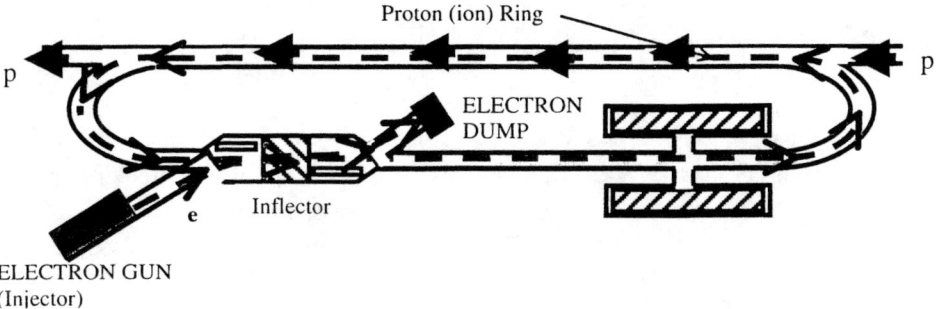

FIGURE 1. Principle Scheme of Electron Cooler with Circulating Electron Beam

Therefore electron temperature in cooling process increase very fast (Fig. 2) and electron beam is to be renewed after electrons have got a significant temperature. One can see reduction of electron heating rate when antiproton temperature decreases

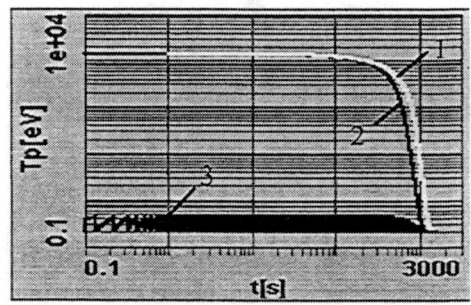

FIGURE 2. Simulation of the cooling process with equations (1), (2) for Recycler parameters: $N_p = 1.4 \cdot 10^{13}$, $N_e = 8 \cdot 10^{12}$, $\eta = 0.02$, $T_p(0) = 1.4$ KeV ($\varepsilon = 3.3$ π·mm·mrad), $T_e = 0.2$ eV, the injection with repetition frequency is 10 Hz. 1 – T_p at cooling with circulating electrons, 2 – T_p at cooling with single pass electrons, 3 – electron temperature.

2.2. Electron beam magnetization

An efficient way of electron cooling rate increase is magnetization of electron beam, i.e. an use of longitudinal magnetic field [6], which transport the electron beam from the gun cathode to collector through drift chambers including cooling section. The criterion of electron beam magnetization has a clear physics meaning: radius of electron Larmor spiral in magnetic field is smaller of distance between electrons [6]:

$$\rho \leq (n_e)^{-1/3} \text{ when } B > B_{min\,1} \equiv \frac{m_I}{e}\sqrt{T_\perp mc^2}\left(\frac{I_e}{\beta\gamma\pi a^2 ec}\right)^{1/3}. \quad (3)$$

Here n_e is electron density in the electron rest frame, I_e – electron beam current, e, m – electron charge and mass, βc – electron velocity, $\gamma = (1-\beta^2)^{-1/2}$. Computer simulations with single pass electron beam show significant, up to several times, growth of cooling rate when $B \Rightarrow B_{min1}$ and its saturation at $B > B_{min1}$.

When electron beam is magnetized, electron drift caused by its own electric and magnetic fields is sufficiently supported and does not exceed electron thermo velocity if

$$B > B_{min\,2} \equiv \frac{2I_e}{\beta\gamma a}\cdot\sqrt{\frac{m}{T_\perp}}. \quad (4)$$

Comparing the criteria (3) and (4) one can see, that $B_{min1} > B_{min2}$, when

$$I < \frac{\beta\gamma ce}{\sqrt{\pi}\,r_e^{3/2}}\cdot\sqrt{a}\left(\frac{T_\perp}{2mc^2}\right)^{3/2} \approx 17.7\,\beta\gamma\sqrt{a_{[cm]}}\cdot(T_\perp)^{3/2}_{[eV]}, \quad (5)$$

where r_e is electron classic radius.

2.3. The "flattened" distribution

of electron plays also an essential role in electron cooling [6]. However in the case of circulating beam one can expect its destruction due to influence of several effects, like interaction with particles to be cooled (electron heating), intrabeam scattering in electron beam etc. This problem is to be studied in experiments.

2.4. Nonuniform heating of the magnetized electron beam

by the particles to be cooled can take place because efficiency of the electrons and particles interaction differs for the electron beam core, where particles have high velocity of betatron oscillations, and for the other part of the electron beam, where this velocity is low.

2.5. Intrabeam scattering in electron beam

causes longitudinal-transverse relaxation, or equalizing of longitudinal and transverse temperatures of the electron beam (see details in Ref. 6). If electron beam is magnetized, the relaxation rate can be estimated as the following:

$$\left(\frac{dT_\parallel}{dt}\right)_{Magnetized} \leq \frac{8\pi}{\gamma^2} \cdot cr_e^2 (n_e)_{lab} (mc^2)^{3/2} \frac{\sqrt{T_\parallel}}{T_\perp}. \qquad (6)$$

2.6. Circulating beam stability

problem is strongly connected with the beam space charge. Normally the focusing structure of the ring described here [7, 8] allows to operate below transition energy (see section 4), which means absence of negative mass instability. Nevertheless, there are other contributors provoking the beam instability, like resistive wall impedance, for instance.

3. LOW ENERGY PARTICLES TOROIDAL ACCUMULATOR (LEPTA)

As a candidate of a storage ring for electron cooling with circulating electron beam the machine described initially in Ref. [7] can be proposed. The Low Energy Particle Toroidal Accelerator (LEPTA) consists of solenoid, which in this case has a form of racetrack with two toroidal sections and two straight ones (Fig. 3). Electron injection is provided using special septum, where electrons displace in horizontal direction and come into the kicker, which moves them vertically (see details in Ref. 8). Single turn injection is used.

4. ELECTRON DYNAMICS IN THE RING

In the ring electron travel along magnetic field lines, formed by homogeneous longitudinal magnetic field and quadrupole one. The field lines form the magnetic surfaces if the quadrupole field gradient G is small enough:

$$G < G_{crit} \equiv 2\pi \frac{qB}{C_{ring}}, \quad \text{or} \quad q > \frac{G}{B} \cdot \frac{C_{ring}}{2\pi}, \qquad (7)$$

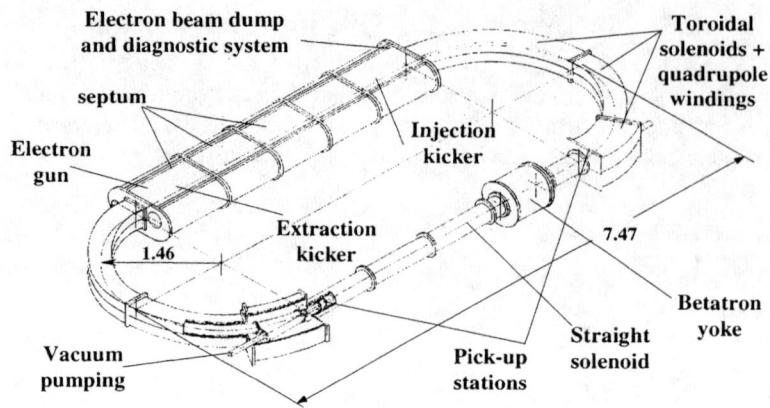

FIGURE 3. The scheme of LEPTA

where q is the number of spiral winding turns per ring circumference C_{ring}, B – longitudinal magnetic field.

Fast Larmor rotation of electron around magnetic field lines occurs with the periodicity $\lambda_L = 2\pi\rho_L$, where ρ_L is electron Larmor radius.

The electron drift along magnetic surfaces is an equivalent of slow two-dimensional oscillations around axis of quadrupole magnetic field: number of such *"betatron" oscillations* per electron revolution Q_{bet} is:

$$Q_{bet} = \frac{1}{2q} \cdot \left(\frac{G}{2\pi B}\right)^2 \cdot C_{ring} \cdot L_{quad}, \qquad (8)$$

where L_{quad} is the summary length of quadrupole field sections. This formula describes the case of short Larmor period: $Q_L \equiv C_{ring}/\lambda_L \gg 1$. The transition energy of the ring corresponds to

$$\gamma_{tr}^2 = \pm Q_L Q_{bet} \cdot \frac{C_{ring}}{2\pi R_s} \qquad (9)$$

where R_s is the equilibrium orbit radius in torus. The sign choice depends on direction of $\vec{B}$ vector. It means one can have $\gamma_{tr}^2 < 0$, i.e. at such a regime $d\omega/dp > 0$ and at any energy no negative mass instability takes place. *Resonance conditions* are similar to those for strong focusing systems:

$$mQ_L + nQ_{bet} = \ell \qquad (10)$$

where ℓ, n, m are integers. Electron transverse velocities (temperature) in the fields of the focusing system are rather large:

$$Q_G \equiv \frac{p_\perp}{p} \sim 2\pi \frac{Ga}{C_{ring}} \sim 0.2. \tag{11}$$

It means that the cooling section must be free of quadrupole field, i.e. the ring has to have a section structure. To provide small Q_G an adiabatic transition from quadrupole field to homogeneous one is necessary. In this case

$$(Q_G)_{adiabatic} \sim \frac{a\ell}{\rho_L^2} \cdot e^{-\ell/\rho_L} \cdot \frac{Ga}{B}, \tag{12}$$

where ℓ is the length of transition region.

5. INJECTORS

Electron acceleration and beam formation has to satisfy the main requirement: the accelerated beam has to have small momentum and angular spreads after injection into the cooler. There are three *candidates:*
1) Electrostatic generator (Van der Graaf or Cockroft-Walton accelerator) in pulsed mode. It needs a large capacity to provide a small energy variation during the pulse. Furthermore, the high voltage problems, like isolation, pulsed load, high pulse current and so on make the task complicated enough.
2) Linac has a significant advantage when both beams – the cooling one and one to be cooled are bunched. To fill the cooler with coasting electron beam, one needs to have a high peak current in the linac. On other hand, to have a low electron momentum spread one should avoid phaze oscillations in the linac. It means the linac has to work at low frequency. Then the problem of cavity quality and RF power consumption appears.
3) Modified betatron described below, looks compatible with cooler scheme presented here.

5.1. The Modified Betatron

has the same focusing system as LEPTA and additional betatron yoke, which generates accelerating voltage V_{accel}. Electron momentum increases in time as

$$p_e(t) = p_{inj} + \frac{e}{C_{cooler}} \int_{t_{inj}}^{t} [V(t)]_{accel} dt. \tag{13}$$

The bending field in toroids has to be varied accordingly to the equation:

$$B_\perp(t) = \frac{p_e(t)c}{eR} = (B_\perp)_{inj} + \frac{ec}{RC_{cooler}} \int_{t_{inj}}^{t} V_{accel} dt, \quad (14)$$

where R is the toroidal radius, C_{cooler} – the ring circumference. Therefore, if $(B_\perp)_{inj}= pc/eR$, then $p(t) = B_\perp(t)/eR$ and the central orbit is stable: it is the equilibrium orbit.

Due to variation of accelerating voltage electrons injected and extracted in different moments get different energy (momentum) in accordance with the equation (13):

$$\Delta p = \frac{e}{C_{cooler}} \left[\int_{t_{inj}+\tau_{inj}}^{t_{extr}+\tau_{extr}} V_{accel} dt - \int_{t_{inj}}^{t_{extr}} V_{accel} dt \right]. \quad (15)$$

Here τ_{inj} and τ_{extr} are electron revolution periods in the ring at injection and extraction. This effect creates the electron beam momentum spread:

$$\frac{\Delta p}{p_{extr}} \approx \frac{e}{p_{extr} c} \left[\frac{V_{extr}}{\beta_{extr}} - \frac{V_{inj}}{\beta_{inj}} \right]. \quad (16)$$

It is a rather small magnitude. For instance, if $p_{extr} = 9\ GeV/c$, $E_{inj} = 10\ keV$, $V_{inj} \approx V_{extr} \approx 1\ kV$, $\Delta p/p_{extr} \approx 5 \cdot 10^{-6}$.

Which experience we have? There are several machines with high current beam in MeV energy range (Table 2). One of them is Modified Betatron of the Naval Research Laboratory [9].

However the beams in these machines have a large electron momentum spread, of the order of 10^{-2} or larger.

The version of the LEPTA storage ring, which makes possible not only circulation of electrons, but their acceleration also is the modified betatron MOBY under manufacturing at JINR [8].

TABLE 2. High current electron accelerators

Accelerator \ Parameter	Electron accelerators of Budker INP			Modified Betatron of the Naval Research Laboratory
	B3[1]	B3M[2]	B4[2]	
Energy, MeV	3	2.1	3	20
Current, A	300	2	20	1000
Circulation time, μs	100	200	100	700

[1] Betatron with spiral storage of electron
[2] Synchrotron with betatron preacceleration

CONCLUSION

1) Electron cooling with circulating electron beam has large advantage comparatively to conventional cooler scheme.
2) The Modified Betatron is the most promising candidate for injector.
3) Realisation of the proposed approach needs further studies.

This work was supported by the Russian Foundation for Basic Research (grant No. 96-02-17211) and INTAS (grant No. 96-0966).

REFERENCES

1. Nagaitsev S., "Electron Cooling for the Fermilab Recycler", in *Proceedings of XVIII International Conference on High Energy Accelerators, Dubna,* 1998, to be published.
2. Blashe K., Dikansky N., Eidelman Yu., Parkhomchuk V., Pestrikov D., Salimov R., Skrinsky A., Strucmeier J., "Electron-Nucleon Collider Design Study (ENC)", *Proc. 6th European Particle Accelerator Conference, Stockholm, 22-26 June,* 1998.
3. Balewski K., Brinkmann R., Derbenev Ya., Floettmann K., Holtkamp N., Schmitz M., Voss G.-A., Wesolovski P., Yeremian D., "Preliminary Study of Electron Cooling Possibility of Hadronic Beam at PETRA", *Proc. 6th European Particle Accelerator Conference, Stockholm, 22-26 June,* 1998, p. 1079.
4. Maier R., Pfister U., Range J., "The COSY Juelich Project April 1991 status", in Proc. 1991 IEEE Part. Accel. Conf. (IEEE cat. No. 91CH3038-7, IEEE Piscataway NJ) p. 2808.
5. Kuksanov N.K., Meshkov I.N., Salimov R.A., Smirnov B.N., Veis M.E. et al., "High voltage electron cooling device", *Proc. 13 Int. Conf. on Accelerators, Novosibirsk,* 1987, p. 348.
6. Meshkov I., *Phys. Part. Nucl.* **25** (6) 631 (1994).
7. Meshkov I., Skrinsky A., *Nucl. Inst. Methods* **A 379** 41 (1996).
8. Korotaev Yu., Meshkov I., Mironov S., Sidorin A., Syresin E., "The modified betatron prototype dedicated to electron cooling", *Proc. 6th European Particle Accelerator Conference, Stockholm, 22-26 June,* 1998, p. 1061.
9. Kapetanacos C., Len L., Smith T., Marsh S., Loschialpo P., Dialetis D., Mathew J., *Phys. Fluids* **B 5** (7), p. 2295.

Transverse Electron-Ion Instability in Ion Storage Rings with High Current

P. R. Zenkevich

Institute for Theoretical and Experimental Physics, B.Cheremushkinskaya, 25, Moscow, Russia

Abstract. A transverse electron-ion instability in ion storage rings with high current is investigated by use of new concept «two-stream transverse impedance», depending on the parameters of the electron beam appeared due to ionization of the residual gas inside the vacuum chamber. A formula connecting this impedance to the space-charge impedance of the ion beam and the neutralization degree has been derived. It is shown that in the ion rings there is «natural» neutralization degree which is defined by an equilibrium between the ionization rate and the electron losses due to heating by electron-ion elastic Coulomb scattering. Application of the theory to an estimation of the «natural» neutralization degree and the threshold momentum spread of the electron-ion instability for ITEP ion storage rings (TWAC) has shown that electron-ion instability could be dangerous for this facility.

INTRODUCTION

A two stream transverse instability of the circular beams with an opposite sign of the particle charge was examined earlier in a lot of papers (Refs. [1-4]). It was shown that the instability has resonant character and may be suppressed by Landau damping if the both beams have frequency spreads. In the present note we investigate a dependence of Landau damping on beam distribution functions by introducing a new concept of «two stream transverse impedance» which can be considered as new kind of one beam transverse impedance. Such method permits to include two stream phenomena in the standard scheme of transverse stability analysis for one beam. By use of this technique a formula for the threshold values of neutralization degree η (η is a relation of the electrical charge of the electrons to the electrical charge of the stored ions) and the ion momentum spread for different electron and ion distribution functions on incoherent frequencies has been found.

Value of η in the ion storage rings is estimated with account of the ionization rate and the electron heating due to Coulomb scattering of the electrons on the circulating ions. It is shown that equilibrium between these two effects results in some «natural» neutralization degree which is calculated for constructed ITEP storage rings.

Comparison of the threshold momentum spread of the ions corresponding this value of neutralization with the threshold momentum spread for one beam transverse instability has shown that electron-ion instability can be dangerous for this facility.

EQUATIONS OF MOTION

Equations of the transverse motion for two individual particles of the circular beams may be written as follows:

$$\begin{cases} \left(\dfrac{\partial}{\partial t}+\Omega\dfrac{\partial}{\partial \theta}\right)^2 x + \Omega_0^2\left(Q^2 x + Q_1^2 x_c - Q_2^2 y_c\right) = 0 \\ \dfrac{d^2 y}{dt^2} + \Omega_0^2\left(q^2 y + q_1^2 y_c - q_2^2 x_c\right) = 0 \end{cases} \quad (1)$$

Here x, y are transverse coordinates of two particles (x corresponds to circulating beam, y – to stored immovable beam), x_c and y_c are coordinates of the beam center of gravity, Ω and Ω_0 are the revolution frequencies of the test particle and the equilibrium particle of the circulating beam, t is a time, θ is the azimuthal angle, Q and q are incoherent «betatron» frequencies for two test particles; Q_1^2, Q_2^2, q_1^2 and q_2^2 are complex parameters which describe electromagnetic forces proportional to coordinates of the beam centers of gravity. Here we assume that revolution frequency Ω and the betatron incoherent frequency of the circulating ions Q depend on some «external» parameter u, and the betatron incoherent frequency of the electrons q depends on the external parameter v (physical sense of these two parameters will be discussed later).

We solve Eqs.(1) assuming that the coordinates of the test particles and the beam centers oscillate harmonically in time and space:

$$x = a\exp[i(n\theta-\omega t)], \quad y = b\exp[i(n\theta-\omega t)]$$
$$x_c = A\exp[i(n\theta-\omega t)], \quad y_c = B\exp[i(n\theta-\omega t)]$$

Substituting these expressions into Eqs.(1) and averaging on incoherent frequencies we obtain:

$$\begin{cases} A(1+Q_1^2 R_1) - BQ_2^2 R_1 = 0 \\ -A q_2^2 R_2 + B(1+q_1^2 R_2) = 0 \end{cases} \quad (2)$$

Here the dispersion integrals $R_{1,2}$ are defined by:

$$R_1 = \int_{-\infty}^{\infty}\dfrac{f(u)du}{Q^2(u)-[\omega-n\Omega(u)]^2/\Omega_0^2}$$

$$R_2 = \int_{-\infty}^{\infty}\dfrac{f(v)dv}{q^2(v)-\omega^2/\Omega_0^2} \quad (3)$$

where $f(u)$ and $f(v)$ are normalized to unity distribution function on u and v.

Taking into account that the determinant of an uniform linear system should be equal to zero, we can write the dispersion equation for slow wave (with $\omega \approx \Omega_0 (n - Q)$) in the following form:

$$(1+\delta Q_1 V_1)(1+\delta q_1 V_2)-\delta^2 V_1 V_2=0 \qquad (4)$$

where δQ_1 and δq_1 are coherent frequency shifts due to species-species forces, δ is the increment of the two-stream instability for monochromatic beams in absence of species-species forces for monochromatic beams. These parameters are defined by:

$$\delta Q_1=Q_1^2/2Q; \quad \delta q_1=q_1^2/2q; \quad \delta^2=Q_2^2 q_2^2/4Qq \qquad (5)$$

Here dispersion integrals V_1 and V_2 are:

$$V_1 = \Omega_0 \int_{-\infty}^{\infty} \frac{f(u)du}{[n-Q(u)]\Omega(u)-\omega}, \quad V_2 = \Omega_0 \int_{-\infty}^{\infty} \frac{f(v)dv}{q(v)\Omega_0 - \omega} \qquad (6)$$

Solution of dispersion equation Eq.(4) can be performed analytically in the simplest cases: monochromatic beams (distribution function on x $f(x)=\delta(x)$); beams with Lorentzian distributions ($f(x)=\Delta/[\pi(x^2+\Delta^2)]$); beams with semicircle distributions ($f(x)=2\Delta/[\pi\sqrt{\Delta^2-x^2}]$). For Lorentzian distribution the result can be written as:

$$v= \omega/\Omega_0=q_0+d-i(\Delta_i+\Delta_e)/2\pm i\sqrt{\delta^2 - \frac{1}{4}[d-i(\Delta_i-\Delta_e)]^2},$$

$$d = (n-Q_0-\delta Q_1-q_0-\delta q_1)/2, \quad \Delta_i = S\frac{\Delta p}{p} \qquad (7)$$

Here $S = n-Q+\xi/\Gamma$ (ξ is the ring chromaticity, Γ is the momentum compaction factor defined by $\Gamma= \gamma_t^{-2}-\gamma^{-2}$), $\Delta p/p$ – momentum spread in the ion beam (here we assume that dispersion of the ions on the betatron and revolution frequencies appears only due to dispersion on momentum), Δ_e – electron spread on incoherent frequency q. Let us underline, that this spread can appear due to two factors: non-linearity of the ion electrical field and due to azimuthal variations of its gradient. Strictly speaking, analysis of the first effect requires a solution of linearized Vlasov's system; however, such solutions have shown Ref.[1,5] that for «good» distribution functions (with negative derivative from stationary distribution function on energy) results of the exact solution qualitatively coincides with results obtained by use of our «phenomenological» equations.

We see from Eq.(7) that the increment reaches its maximal value if $d=0$, i.e.

$$n = Q_0+\delta Q_1+q_0+\delta q_1 \qquad (8)$$

This condition has simple physical sense: sum of beam coherent frequencies calculated with account of one beam coherent shifts should be equal to integer number. For resonance conditions a relation of coherent amplitudes is defined by

$$\frac{a_i}{a_e} = \frac{[\delta Q_1 + i(\alpha - \Delta_i)]\delta Q_{1,2}}{[\delta q_1 + i(\alpha - \Delta_e)]i(\alpha - \Delta_i)} \qquad (9)$$

Here $\alpha = \mathrm{Im}\,v$, $\delta Q_{1,2} = Q_2^2/2Q$ is the coherent frequency shift of the coasting beam due to two stream coherent forces. Investigation of Eq.(9) shows that for fast instability ($\alpha \gg \Delta_i$) and typical values of machine parameters, as a rule, $|a_i|/|a_e| \ll 1$ (as an example, for ITEP ion complex parameters, $q=70$, $\delta Q_1=0.2$, $\eta=0.1$ and $\gamma=1.7$, we find that $|a_i|/|a_e|=0.002$. However, if $\alpha \cong \Delta_i$, $|a_i|/|a_e|$ goes to infinity. Investigation of a dependence $|a_i|/|a_e|$ on the beam parameters for more realistic distribution functions shows that near the instability threshold value $|a_i|/|a_e|$ can be an order of unity. Thus, in dependence on beam history and machine parameters, two stream instability can result not only in the ion heating, but, as well, in the ion losses.

Now let us consider the beam behavior in the stable region. Stability of the solution Eq.(7) requires the spreads which satisfy the following condition:

$$\Delta_i \Delta_e \geq \delta^2 [1+(\frac{2d}{\Delta_e + \Delta_i})^2]^{-1} \tag{10}$$

For $d = 0$ (exact resonance) Eq.(10) provides a simple sufficient condition of the beam stability:

$$\Delta_i \Delta_e \geq \delta^2 \tag{11}$$

Let us underline that in a frame of this approximation (Lorentzian distribution functions) the stability condition does not depend on one-beam coherent frequency shifts δQ_1 and δq_1 (this result occurs due to independence of Landau damping rate on frequencies for Lorentzian distribution). For a case of more realistic semicircular distribution stability criterion transforms to the following form:

$$(\Delta_i^2 - \delta Q_1^2)(\Delta_e^2 - \delta q_1^2) \geq \delta^4 \tag{12}$$

It is clear that for stability we should require, that

$$\Delta_i \geq \delta Q_1, \Delta_e \geq \delta q_1 \tag{13}$$

For arbitrary distribution functions it is convenient to make analysis by introducing new concept of «two stream transverse impedance».

TWO STREAM TRANSVERSE IMPEDANCE

Let us rewrite our dispersion equation in the following form:

$$1 + V_1(\delta Q_1 + \delta Q^{1,2}) = 0 \tag{14}$$

where

$$\delta Q^{1,2} = \delta^2 (\delta q_1 + V_2^{-1})^{-1} \tag{15}$$

It is well known that for one beam the coherent frequency shift corresponds to the transverse impedance $Z_\perp$. Therefore we can write:

$$Z_\perp^{1,2} = Z_\perp^1 \frac{\delta Q^{1,2}}{\delta Q_1} = Z_\perp^1 \delta^2 /[\delta q_1(\delta q_1 + V_2^{-1})] \tag{16}$$

Let us consider in more details an instability due to an interaction between the coasting ion beam and the stored ionization electrons, which is especially dangerous for heavy ion beams with high current (such rings are assumed to use for plasma physics and accelerator experiments). A final aim of these experiments is the advanced studies concerned Heavy Ion Fusion (HIF). In this case the «space charge» transverse impedance of the ion beam Z_{sc} is usually much more than the last terms. This impedance is defined by:

$$Z_\perp^{sc} = iZ_0 R \frac{1}{\beta^2 \gamma^2}(h^{-2} - b^{-2}) \tag{17}$$

Here Z_0 is the free space impedance ($Z_0 = 377$ Ohms), R is the ring radius, h is the beam size and b is the vacuum chamber radius.

Two-stream transverse impedance $Z_\perp^{1,2}$ depends on the electron dispersion integral V_2 which can be written in the following form:

$$V_2 = a(U_0 + iV_0) \Delta q^{-1} \qquad (18)$$

where a is the coefficient depending on the electron distribution function, Δq is a half width of the corresponding distribution function at its half height, U_0 and V_0 is are defined by well known formulae:

$$U_0 = PV \int_{-\infty}^{\infty} \frac{\varphi(x)dx}{x - \alpha}, \quad V_0 = i\pi\varphi(\alpha) \qquad (19)$$

Here function $\varphi(x)$ is chosen such a way that $\varphi(1) = 1$.

Let us introduce $Z_a = i(U_0 + iV_0)^{-1}$. Using Eq.(17)-(19) and given in Appendix expressions for frequencies, we obtain:

$$Z_\perp^{1,2} = Z_\perp^{sc} \gamma^2 ak / (ak - \text{Im} Z_a + i \, \text{Re} Z_a) \qquad (20)$$

where parameter k is connected with physical parameters of the machine by

$$k = \eta(1 - a^2/b^2)/[2(1-\eta)\frac{\Delta q}{q}] \qquad (21)$$

When $ak = \text{Im} Z_a$, two stream impedance reaches its maximal pure active value

$$Z_\perp^{1,2} = -iZ_\perp^{sc} \gamma^2 \Lambda(k) \qquad (22)$$

where

$$\Lambda(k) = ak/\text{Re} Z_a \qquad (23)$$

From Eq.(20) we see that k_0 (a maximal value of k) is equal to $(\text{Im} Z_a)_{max}/a$. Value of this coefficient depends only on a form of the distribution function. Then, using Eq.(21), we find out, that the maximal neutralization degree

$$\eta_{max} = 2\frac{\Delta q}{q} k_0 (1 - a^2/b^2 + 2\frac{\Delta q}{q} k_0)^{-1} \qquad (24)$$

This limitation appears due to proportionality between the coherent shift of the electron frequency and the neutralization degree η (for values of $\eta \geq \eta_{max}$ Landau damping of the electron oscillations is absent), and electron-ion oscillations are unstable. For given frequency spreads of electrons and ions there is a threshold value of η ($\eta = \eta_{thr}$) which determines the stability limit. In order to find η_{thr} it is necessary to find dependence Λ on k.

This dependence is plotted in Fig.1 for three different distribution functions $\varphi(x)$: 1) Gaussian function $\varphi(x) = \exp(-x^2/2)$ ($a = 0.470$, $k = 4.055$); 2) quartic function $\varphi_{(x)} = (1-x^2)^2$ ($a = 0.507$, $k = 1.315$) 3) $\varphi_{(x)} = 1/\sqrt{1-x^2}$ ($a = 0.55$, $k = 0.578$).

We see from the figure that for small values of k (and, correspondingly, small values of the neutralization degree) the two stream impedance weakly depends on a form of the electron distribution function; but this dependence becomes stronger if η approaches to its maximal value η_{max}. Physical considerations show that real electron distribution should have long tails and therefore in the following analysis we will consider quartic and Gaussian distribution of the electrons.

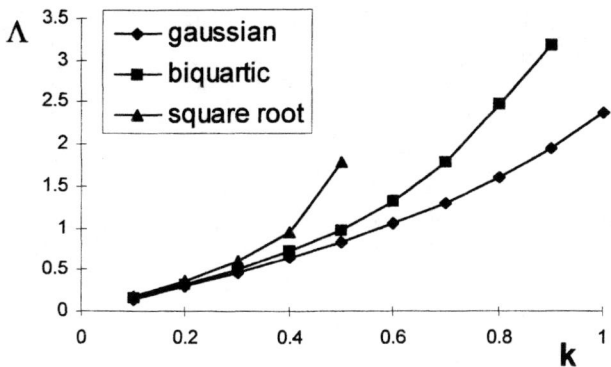

FIGURE 1. Dependence of Λ on k for different distribution functions.

This theory permits to calculate two-stream transverse impedance which in the resonance point has purely active character; using this impedance, we can find the stability conditions for the ion coasting beam by well known standard procedure (see, for example, Ref.[5]). If we neglect all other sources of impedance, we can write (due to a proportionality of this two-stream transverse impedance to the ion beam space charge impedance) the following condition for the threshold momentum spread of the ion beam $(\frac{\Delta p}{p})_{thr}$ (here $\frac{\Delta p}{p}$ is a half width at half height of the distribution function on momentum deviation):

$$(\frac{\Delta p}{p})_{thr} = 0.125\, I\, \frac{R|Z_{\perp}^{sc}|}{U_p}\, \frac{Z_i}{A}\, \frac{1}{\beta\gamma Q|\Gamma|}\, \frac{1}{(q + \xi/\Gamma) F(\gamma^2 \Lambda)} \quad (25)$$

In Eq.(25) U_p is the proton «rest voltage», β and γ – relativistic parameters, I – the ion current, Z_i and A – charge and atomic number of the ion, and factor F depends on the two-stream transverse impedance which is proportional to $\gamma^2\Lambda$; this function is plotted in Fig.2 for different ion distributions.

We see from the figure, that for small $\gamma^2\Lambda$ value of F strongly depends on the chosen ion distribution function; for big values of $\gamma^2\Lambda$ for all distributions $F \cong 1/\sqrt{1 + (\gamma^2\Lambda)^2}$. Using these figures we can find η_{thr} for given machine parameters. For example, let us assume that the momentum spread is determined by one beam stability conditions for the lowest transverse mode; then the value of factor F corresponding the threshold of two beam instability ($F^{e,i}$) may be found from the simple condition:

$$F^{e,i} = F^1 S^1 / S^{e,i} \quad (26)$$

Here values of factors F^1 and S^1 corresponds to a threshold of one beam instability, $S_{e,i} = q + \xi/\Gamma$.

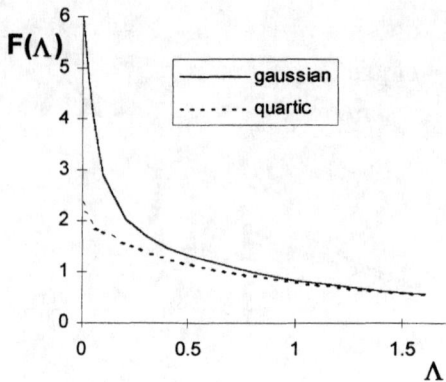

FIGURE 2. Dependence of F on Λ for different distribution functions.

ESTIMATION OF «NATURAL» NEUTRALIZATION DEGREE

We see that the threshold momentum spread for given beam current depends on the neutralization degree. Let us estimate value of the neutralization degree for «natural» conditions (clearing electrodes are absent). Equation describing a dependence of the electron number N_e on time is:

$$\frac{dN_e}{dt} = -\frac{N_e}{\tau_l} + \frac{N_i Z_i}{\tau_{neutr}} \tag{27}$$

Here τ_{life} is the electron life-time, τ_{neutr} is the time of space charge neutralization. The stationary (maximal) neutralization degree $\eta = \tau_{life}/\tau_{neutr}$. The electron life time is determined by balance between heating and cooling processes. Main source of the electron heating is their Coulomb scattering on circulating ions; the corresponding heating rate is defined by:

$$\frac{dW_e}{dt} = \frac{4\pi n_i Z_i^2 e^4}{m_e \beta c} L \tag{28}$$

where L is the Coulomb logarithm (for further we assume $L=20$).

The electron will be lost if its energy is more than a depth of the ion beam potential well W, which is defined by

$$W = \pi n_i r_e Z_i a^2 (1-\eta) \tag{29}$$

From these equations we find:

$$\tau_{eife} = \frac{a^2 \beta}{4 r_e Z_i c L}(1-\eta) \tag{30}$$

Neutralization time is defined by:

$$\tau_{neutr} = \frac{Z_i}{d_m \sigma_{ion} \beta c} \tag{31}$$

Here d_m is the density of the residual gas in the vacuum chamber, σ_{ion} is the ionization cross-section Ref.[4].

$$\sigma_{ion} = Z_i^2 K \beta^2 [C + M^2 (\ln \frac{\beta^2}{1-\beta^2} - \beta^2)]$$

In this equation $K = 1.87 \cdot 10^{-24}$ m^2, coefficients C and M^2 depend on the kind of residual gas in the vacuum chamber (for N_2: $C = 34.8$ and $M^2 = 3.7$). Molecular density in vacuum chamber d_m (in m^{-3}) is expressed through pressure P_m (in 10^{-10} Torr) by:

$$d_m = 3.3 \cdot 10^{12} P_m$$

Using these formulae, we obtain:

$$\eta = \eta_0/(1+\eta_0), \qquad \eta_0 = \frac{a^2 \beta^2 \sigma_{ion} d_m}{4 r_e Z_i^2} \frac{1}{L} \qquad (32)$$

Substituting all constants, we find the final simple expression for η_0:

$$\eta_0 = 2.8 \times 10^{-3} P_m \, a^2 \, C \, \Phi(\beta) \qquad (33)$$

Here a in cm, and factor $\Phi(\beta)$ ($\Phi(\beta) \approx 1$) is defined by:

$$\Phi(\beta) = 1 + \frac{M^2}{C} (\ln \frac{\beta^2}{1-\beta^2} - \beta^2)$$

NUMERICAL APPLICATION

Let us apply this theory to the TWAC project (ITEP, Moscow, Russia). The list of its parameters is given in Table 1.

Using these data and assuming that $\Delta q/q = 0.1$, we find $\eta_{max} = 0.485$ for Gaussian ion distribution and $\eta_{max} = 0.234$ for quartic ion distribution.

TABLE 1. Parameters of TWAC complex.

Facility	TWAC
Kind of ions	Co_{59}^{+27}
Kinetic energy (MeV/u)	677
β	0.80
γ	1,7
Ring radius R (m)	40
Betatron frequency Q	9.7
Chromaticity	-20
Chamber radius (cm)	4
Horizontal beam emittance (mm·mrad)	80
Vertical beam emittance (mm·mrad)	50
$Z_\perp^{sc}$ (MΩ/m)	21.2
Γ	0.335
Horizontal beam size (cm)	1.8
Vertical beam size (cm)	1.4
I_0 (A)	50
q_0	282

Substituting values of TWAC parameters (kind of the ions Co_{59}^{+27}, $P_m = 10^{-10}$ Torr, $a = 3$ cm, $\beta = 0.8$) and assuming that the residual gas in the vacuum chamber consists from molecular nitrogen, we obtain that for TWAC $\eta_{nat} = 0.22$. Thus, we see that for quartic distribution of the ions beam is unstable.

For Gaussian distributions of the ions $k = 1.2$, $\Lambda(k) = 5.8$, $F(\gamma^2 \Lambda) = 0.06$. Taking into account that in our case $F_1 = 5$ (this value corresponds to small reactive impedance for the most dangerous first harmonics), $S^1=30$ and $S^{e,i}=310$, we find, that in this case $(\frac{\Delta p}{p})_{thr}^{e,i} = 8 (\frac{\Delta p}{p})_{thr}^{i,i}$. Thus, we see that the electron-ion dipole instability can be very dangerous for TWAC facility.

Corresponding estimation for SIS (GSI, Darmstadt) has shown that for SIS the effect is not so dangerous. However, it is useful to mention that if the small (due to electron cooling system) momentum spread in SIS will be less than $(\frac{\Delta p}{p})_{thr}^{e,i}$ the transverse damping system should work in the frequency range of the two-stream instability.

SUMMARY AND DISCUSSION.

1. It is convenient to analyze two-stream dipole instability by use of new concept of «two-stream transverse impedance» $Z_\perp^{1,2}$ which permits to include two-stream phenomenon in the standard scheme of analysis of one-beam transverse stability for the coasting beam.

2. For space charge dominated transverse impedance of the ions the electron-ion oscillations can be stable only if the beam neutralization degree satisfies the condition $\eta < \eta_{max}$, where η_{max} depends on the electron distribution on incoherent frequencies (η_{max} increases for long tail distributions) and relative electron spread on incoherent frequencies.

3. For $\eta < \eta_{max}$, electron-ion oscillations are stable only if the ion beam momentum spread is more than some threshold value defined by machine parameters and the ion and electron distribution functions.

4. In the ion storage rings there is equilibrium neutralization degree η_{nat}, which is defined by balance between the rate of the electron losses due to Coulomb scattering of the electrons by the circulating ions and the ionization rate.

5. Numerical estimations have shown that for constructed ITEP complex TWAC electron-ion instability can be more dangerous than one beam transverse instability due to moderate ring relativism (increasing a value of two-beam forces comparatively with one beam interaction) and comparatively low pressure.

In a case of the instability appearance it is necessary to use for its suppression the well known methods, such as beam clearing, beam shaking., transverse damping system and etc.

Let us remark, that the theory has semi-qualitative character, since a lot of effects is omitted in a frame of accepted simplifying assumptions. For example, we have not taken into account an influence of the magnetic field of the accelerator on electron dynamics. Nevertheless we guess that this simple physical considerations are useful to understand mechanism of the electron-ion instability in high current ion storage rings.

APPENDIX

For space charge dominated beam it is convenient to express all ion parameters through the ion incoherent frequency shift δQ_{inc}, which is defined by

$$\delta Q_{inc} = \frac{r_p N_i Z_i^2 R}{\pi \beta^2 \gamma^3 A Q a_z (a_z + a_x)}$$

Here Z_i and A are the charge and atomic ion number, x, z are transverse coordinates, a_x and a_z are the corresponding beam sizes; parameter δQ_{inc} is given for vertical oscillations. Using (Eq.16), we can write:

$$Q^2 = 2\delta Q_{inc} Q_0 (1 - a^2/b^2)$$
$$Q_2^2 = 2\delta Q_{inc} Q_0 \, \eta \gamma^2 (1 - a^2/b^2)$$

It is convenient to express all electron parameters through electron incoherent frequency in the ion focusing field q_e, which is defined by

$$q_e^2 = 2 r_e N_i Z_i R / [\pi \beta^2 \, a_z (a_x + a_z)]$$

Then

$$q^2 = q_e^2 (1 - \eta)$$
$$q_1^2 = \eta q_e^2 (1 - a^2/b^2)$$
$$q_2^2 = q_e^2 (1 - a^2/b^2)$$

Incoherent frequency shift of the ions δQ_{inc} and the squared incoherent frequency of the electrons q_e^2 are connected by the following expression:

$$q_e^2 = 2 \, Q \, \delta Q_{inc} \, \frac{m_p A \gamma^2}{m_e Z_i}$$

Sometimes it is more convenient to use the ion current I instead of number of the ions N_i, which are connected by:

$$I = Z_i \, e \, \beta c N_i / 2\pi R$$

ACKNOWLEDGMENTS

The author is very grateful to I.Hofmann (GSI) for an interest to the topic and the useful discussions, to D.Koshkarev (ITEP) for useful remarks and to E.Mustafin (ITEP) for help with numerical calculations.

REFERENCES

1. P.R.Zenkevich and D.G.Koshkarev, «Coupling Resonances of the transverse oscillations of two circular beams», reprint ITEP -1060, (1970); *Particle accelerators,* 3 (1972), p.1.
2. E.Keil, B.Zotter, «Landau Damping of Coupled Electron-Proton Oscillations», CERN-ISR-TH/71-58, (1971).
3. L.J.Laslett, A.M.Sessler and D.Moehl, «Transverse two-stream instability in the presence of strong species-species and image forces», *Nuclear Instruments and Methods,* 121 (1974), p.517-524.
4. Baconnier, A.Poncet, and R.F.Tavarez, «Neutralization of accelerator beams by ionization of the residual gas», in *CAS Proceedings,* CAS 94-01,p.525.
5. B.Zotter and F.Zacherer, «Transverse Instabilities of Relativistic Particles Beams in Accelerators and Storage Rings», in *CAS Proceedings,* CAS, CERN 77-13, p.175, Geneva, (1987).

High Current Induction Linacs at JINR and Perspective of Their Application for Acceleration of Ions

G.V. Dolbilov

Joint Institute for Nuclear Research, 141980 Dubna, Moscow Region, Russia

Abstract. This report presents a review of a research activity at JINR on construction of nanosecond high current linear induction accelerators (LIA). The problems of heavy ions acceleration in LIA are discussed.

INTRODUCTION

Linear induction accelerators at the Joint Institute for Nuclear Research (JINR) have been developed in the framework of program of collective method of acceleration. Several accelerator complexes have been constructed allowing to produce relativistic electron rings, to fill them with multi-charged ions and to accelerate electron rings [1–6] (see Table 1). These accelerators are used now for variety of another applications, such as free electron lasers, relativistic klystron and two-beam acceleration.

DESIGN FEATURE OF THE LIA AT JINR

The first linear induction accelerator, LIA-3000, has been designed and manufactured by the Efremov Electrotechnical Institute (S. Peterburg) and installed at JINR in 1966. The LIA-3000 accelerator is a 250 ns pulse duration machine with one-step modulator designed in a standard way. Its modulator consists of a forming line and a switch. The forming lines are used as energy storage, and powerful hydrogen thyratrons serve as commutators. High voltage pulses on the accelerating modules are driven by the forming lines. One modulator is used to drive one 50 keV accelerating module consisting of 3 permalloy core. Maximal operating current of electrons is about of 250 A electron beam. The average accelerating gradient of the LIA-3000 is 0.25 MeV/m. This accelerator is in permanent operation for more than thirty years. It has been used in the program of the collective method of acceleration and now is used for the FEL and relativistic klystron research.

TABLE 1. Survey of JINR Induction Accelerators

Name of Accelerator	Date of design	E MeV	Core material	τ ns	f Hz	Cathode types	I_b A
LIA-3000	1966-67	1.5	permalloy	250	1	BaO	300
SILUND	1971-73	1.7	ferrite	15	1	plasma	700
SILUND-2	1977-78	0.8	ferrite	20	50	plasma	1000
SILUND-10	1978-80	0.25	ferrite	20	1	-	8000
SILUND-20	1981-82	2.0	ferrite	20	50	plasma	1000
		2.5				explosive	600
LUEK-20	1985-85	10kV/cm	permalloy	60	20	-	rings
LUEK-20 module	1988-91	1.5		60	1	explosive	1500

FIGURE 1. Accelerator SILUND-2: eU=0.8MeV, I_b=1kA, E=0.5MV/m. (Modulator: U_{out}=17kV, I_{out}=70kA. Thyratron: U_{in}=34kV, I_{in}=4kA.)

Investigations of the collective method of acceleration performed at LIA-3000 accelerator have shown that optimal LIA for acceleration of electron rings should have 15–20 ns pulse duration with current up to 1 kA. To reach this goal, we constructed several accelerators named as SILUND, SILUND-2, SILUND-10 and SILUND-20. These accelerators are designed using accelerating modules with ferrite core. Modulators of these accelerators compress electromagnetic energy in time, thus increasing peak output power. The main feature of our approach consists in using powerful, relatively low-voltage generators (17–50 kV), low-resistance load ($\sim 0.5\Omega$) and a high value of peak power (0.6–5 GW). The modulator operates as follows. The energy is stored in capacitor banks and the hydrogen thyratrons operating in μs pulsed mode are used as primary switches. The power amplification takes place in

FIGURE 2. Accelerator SILUND-20: eU=2.0MeV, I_b=1kA, E=0.5MV/m. (Modulator: U_{out}=17kV, I_{out}=35kA. Thyratron: U_{in}=34kV, I_{in}=2kA.)

FIGURE 3. Accelerator SILUND-10: U=0.25MV, I_b=8kA, E=0.66MV/m. (Modulator: U_{out}=42kV, I_{out}=60kA. Thyratron: U_{in}=45kV, I_{in}=2.5kA.)

chains consisting of capacitors and nonlinear reactors (or nonlinear ferromagnetic lines with distributed parameters) as commutators [7]. As a result, the primary pulses of microsecond duration are transformed into the pulses of nanosecond duration with the higher peak power.

The SILUND-2 accelerator was the first one of the new generation of LIA using compression energy in time. It served as a prototype of the basic accelerator SILUND-20. One modulator of the SILUND-2 linac drives two 0.4 MeV accel-

FIGURE 4. Accelerator LUEK-20: E=1Mv/m. (Modulator: U_{out}=50kV, I_{out}=100kA. Thyratron: U_{in}=45kV, I_{in}=6kA.)

erating modules loaded by 1000 A electron beam. The average gradient of the SILUND-2 and SILUND-20 accelerators is equal to 0.5 MV/m. Layouts the of SILUND-2 and SILUND-20 accelerators are shown in Figs. 1 and 2.

Successful operation of the SILUND-2 and SILUND-20 accelerators have proven the idea of power compression. During construction of the next accelerator, SILUND-10, we used the same ideas, but performed optimization of all the systems in order to increase the power transferred into the beam by means of decreasing of the leakage inductances and impedance matching of all the elements of the scheme. This problem was solved by using of a row of short nonlinear ferromagnetic lines with distributed parameters as commutators. The lines are connected in a series in the preliminary compression chain and in parallel — in the final forming chain. All connection including the lines to the accelerating gaps, are manufactured as matched feeders. One modulator feeds 250 kV induction section at the loading equivalent to the electron beam with current 8000 A. The layout of the 0.25 MV accelerating module of the SILUND-10 accelerator is shown in Fig. 3. Total length of the model is equal 38 cm.

The LUEK-20 accelerator (see Fig. 4) is another accelerator of this generation [5,8]. It has been designed for acceleration of electron rings filled with heavy ions. Modulator of the LUEK-20 accelerator has the following parameters: peak power is 5 GW, voltage on 0.5 Ω load is 50 kV, peak current is 100 kA, pulse duration is 60 ns and repetition rate is 20 Hz. The accelerating gradient of the LUEK-20 accelerator is about of 1 MV/m.

PARAMETERS OF THE BEAMS

Axial guide magnetic field is used for beam focusing in all JINR linear induction accelerators. As a result, two general problems caused by such a focusing take place. The first one is that of suppression of the beam drift off axis. The second problem is that normalized emittance of the beam grows in the process of acceleration.

The drift of the electron beam off axis takes place when geometrical and magnetic axes of the focusing solenoid do not coincide. As a rule, finite value of the transverse component of the magnetic field occurs at the beam axis. Even small value of this field, $B_{tr} \sim 2-4$ Gs, is sufficient to provide transverse drift of the beam $\Delta r \sim 10-20$ mm at the accelerator length about of ~ 5 m. Therefore, special correction of the magnetic field is required to avoid the drift of the electron beam off axis.

Growth of the normalized emittance takes place due to nonlinearities of the focusing fields and of the space charge fields. Fig. 5 presents the results of emittance measurements at the JINR accelerators. Several electron sources have been used: with the hot BaO, plasma and explosive (graphite and graphite-fibre) cathodes. Our investigations have shown that parameters of the electron beam do not depend significantly on the cathode type used. On other hand, there is general tendency that the normalized emittance grows during the acceleration. It is seen also from Fig. 5 that the process of emittance dilution is suppressed at higher energy.

One more peculiar feature of the beams is that the phase density of the particles and the beam brightness ($B_n = I/(\pi \varepsilon_n)^2$) of the central core of the beam are

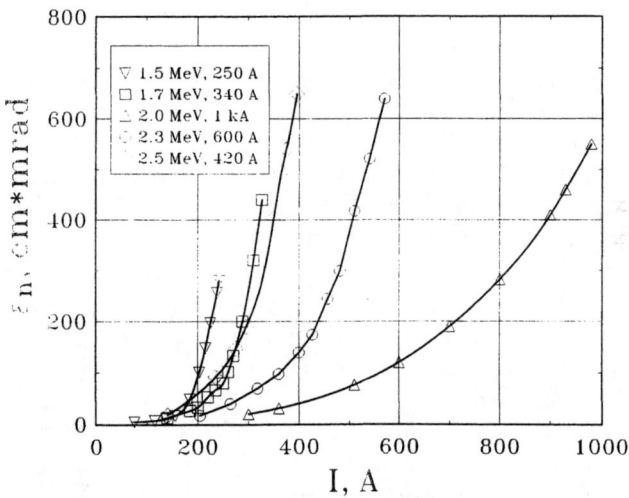

FIGURE 5. Normalized emittance of the central core of the beam versus the beam current of this core.

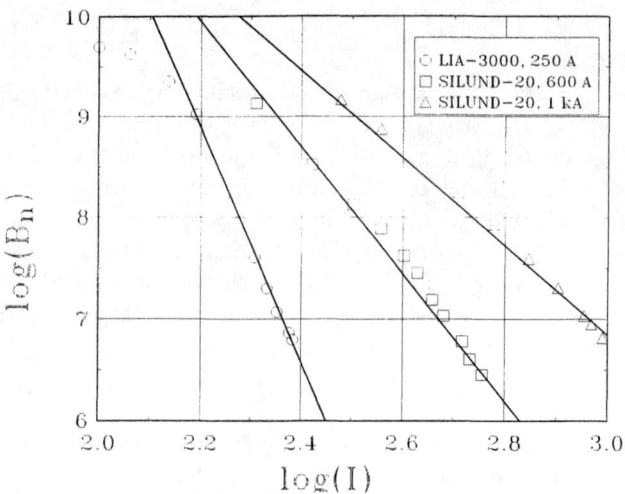

FIGURE 6. Beam brightness of the central core of the beam versus the beam current of this core.

increased with the increase of the total beam current. In Fig. 6 we present the corresponding plot for normalized brightness versus the beam current in the central core for several accelerators with different beam current.

TWO BEAM INDUCTION LINEAR ACCELERATOR

The program of collective method of acceleration at JINR has been stopped in the end of 80-th. Main reason of such a solution has been connected with the rapid progress of traditional accelerating techniques allowing to solve similar problem in a cheaper way. At present linear induction accelerators are used for experimental investigations in the field of relativistic klystrons [9,10] free electron lasers [11,12] and two beam acceleration methods [13,14]. Upon successful completion of the relativistic klystron research program in Dubna [9,10] we search for new fields for LIA applications.

In this paper we discuss a new scheme of two-beam induction accelerator which can be used for acceleration of electrons and heavy ions. The main feature of the proposed accelerator is a novel geometry of accelerating electrodes. They are designed as rf-cavities and are excited by a prebunched driving electron beam (see Fig. 7). Prebunched electron beam can be produced by a gridded electron gun or klystron buncher. Energy losses of the driving beam are recovered by accelerating field in the induction sections. Ions (or electrons) of main beam are accelerated in a strong rf-fields induced in the cavities.

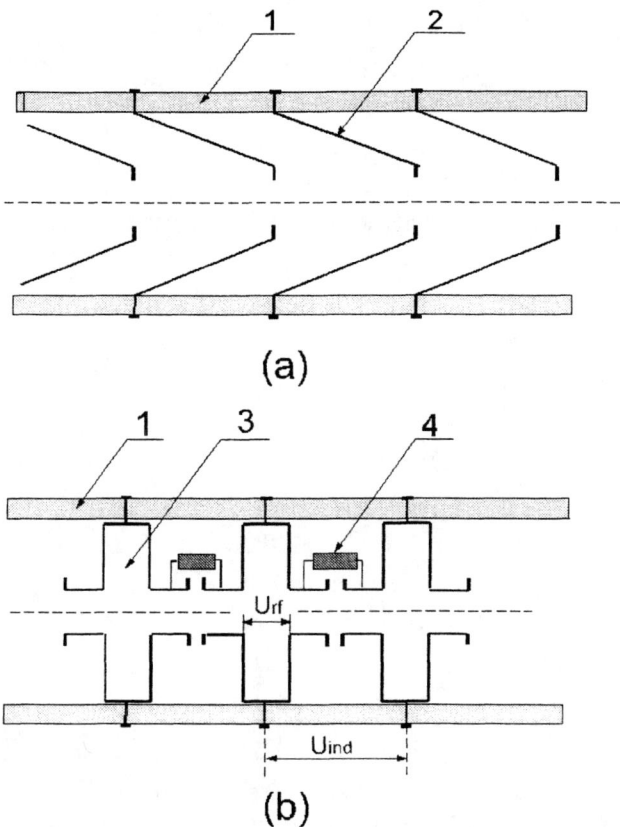

FIGURE 7. Scheme of the accelerating tube: (a) – conventional induction linear accelerator; (b): – two beam induction accelerator. Here: 1 - ceramic tube, 2 - accelerating electrode-diaphragm, 3 - rf cavity, 4 - rf absorbing resistance.

The frequencies of unloaded cavities are slightly larger than the buncher frequency, so an impedance of the cavities is inductive and the phase of the induced voltage of the cavities is close to $\pi/2$. This fact ensures longitudinal stability of the bunched electron beam. Synchronism for the main beam is provided by an appropriate spacing of the rf-structures. To avoid excitation of parasitic modes of oscillations we assume to use technique of resistive insertions [9,10].

Ions Phase Stability

Let the period of the electromagnetic structure be L and the voltage induced in n th cavity be equal to

$$U_n = U_0 \sin[\omega(t - t_0) + \varphi] .$$

The phase delay of the voltage induced in the n+1 th cavity is

$$\Delta\varphi_e = \omega\Delta t = 2\pi \frac{L}{\beta_e \lambda} .$$

The ions-transit time for one period of the structure is equal to $\Delta t = L/v_i$ corresponding to

$$\Delta\varphi_i = \omega\Delta t = 2\pi \frac{L}{\beta_i \lambda} .$$

The synchronism condition takes place when

$$\Delta\varphi_e - \Delta\varphi_i = \pm 2\pi k ,$$

$$\frac{1}{\beta_e} - \frac{1}{\beta_i} = \pm k \frac{\lambda}{L} ,$$

where k is integer. If the relative kinetic energy of the electrons is equal to the relative kinetic energy of the ions

$$\frac{eU_e(z)}{mc^2} = \frac{Z}{A} \frac{eU_i(z)}{Mc^2} ,$$

then $\beta_i = \beta_e$ and $k = 0$. It may be realized in practice when $eU_e \geq 0.511$ MeV and $eU_i \geq 1$ GeV. When $\beta_i \ll \beta_e$, the synchronism condition of the wave and ion is

$$\frac{1}{\beta_i} \simeq k \frac{\lambda}{L} .$$

Synchronism is possible when the electron and ion beams move in opposite directions. In this case

$$\frac{1}{\beta_i} + \frac{1}{\beta_e} = k \frac{\lambda}{L} .$$

The Cavity Excitation

The current of the bunched electron beam is a sum of rf harmonics multiple to the buncher frequency,

$$J(t) = J_e + \sum_{n=1}^{\infty} J_n \cos n\omega t .$$

When the frequency of the unloaded cavity is close to the first harmonic of the buncher frequency, complex amplitude of the induced rf voltage is equal to [15]:

$$U_{rf} = \frac{1}{Y} J_1 M_e ,$$

where Y is complex conductance of the cavity [16]:

$$Y = \frac{\pi}{4Q Z_0}(1 + j\xi) .$$

Interaction factor of the electron and the cavity is [15,17]:

$$M_e \simeq \frac{\sin \frac{\Theta}{2}}{\frac{\Theta}{2}} .$$

Relative detuning of the cavity is

$$\xi = Q \frac{2\Delta f}{f} .$$

where J_1 is first harmonic of the bunched electron beam. Q is quality factor of the cavity. Θ is electron-transit angle and Z_0 is the wave impedance of the cavity. The phase delay of the voltage induced by the driving electron beam is

$$\tan \varphi = -\xi .$$

The modulus of the induced voltage is equal to

$$U_{rf} = \frac{4}{\pi} Q Z_0 J_1 M_e \frac{1}{\sqrt{1+\xi^2}} .$$

The power loss due to cavity excitation is

$$P = \frac{1}{2} U_{rf} J_1 M_e \cos \varphi .$$

Equivalent decelerating voltage for the electrons is

$$U_e = U_{rf} \cos \varphi = \frac{U_{rf}}{\sqrt{1+\xi^2}} .$$

The phase delay must be close to $\pi/2$ and $\xi \gg 1$ in order to excite strong electric field by an electron beam with a relatively low kinetic energy.

The Ion Acceleration

Possible schemes of electrodynamic structure of two-beam induction accelerator are shown in Fig. 8. The structure shown on Fig. 8a has simple geometry and its parameters can be expressed analytically. The second structure (see Fig. 8b) has a higher (by a factor of 2) input impedance and a higher voltage induced by the driving beam. The third variant of the electrodynamic structure (see Fig. 8c) provides larger (by a factor of two) resonance wavelength. This allows one to use a lower (by a factor of 2) ion velocity and a lower (by a factor of 4) ion energy which is important at the initial stage of acceleration.

Let us consider electrodynamic structure presented in Fig. 8a. The voltage on the cavity induced by the bunched electron beam is equal to

$$U_{rf} = \frac{C}{\pi^2} U_e \frac{J_1}{J_i} \frac{a}{d} \frac{M_e}{\cos \varphi_s} , \qquad (1)$$

where J_1 is the first harmonic of the current of the bunched electron beam. J_i is the average current of the ion beam. φ_s is the synchronous phase of the ions. a is the radius of the cavity. d is the diameter of drift tube and constant C is $C = j_{01} J_1^2(j_{01}) = 0.65$. The decelerating voltage U_e is equal to the accelerating voltage U_{ind} of the induction linac, $U_{ind} = E_{ind} L$ (E_{ind} is average induction field). Equation (1) is correct when the quality factor of the cavity is determined by the ion beam load only.

The rate of the energy change by the ion beam in the field of travelling wave of the electrodynamic structure is equal to [17]:

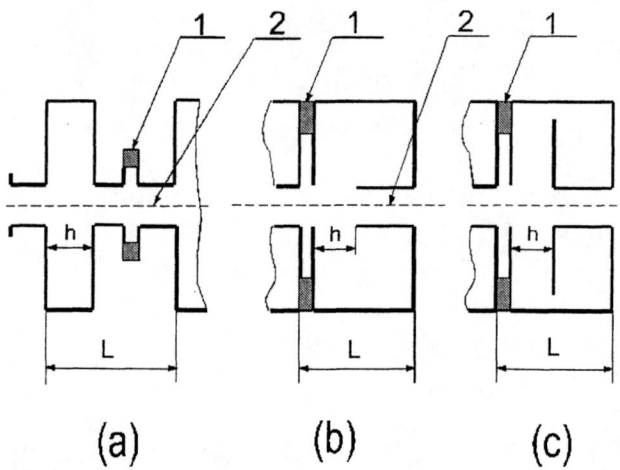

FIGURE 8. Scheme of the possible variants of the two beam accelerator cavities. Here: 1 - rf absorbing insertions, 2 - drift tubes

$$\frac{\Delta U_i}{\Delta z} = \epsilon Z \frac{U_{rf}}{L} \cos\varphi_s M_i ,$$

where interaction factor of the ion and the cavity is

$$M_i \simeq \frac{\sin\frac{\Theta_i}{2}}{\frac{\Theta_i}{2}} ,$$

Θ_i is the ion-transit angle and Z the charge of the ion. For the case of intensive ion beam, when the quality factor of the loaded cavity is $Q_{load} = Q_{ion}$, we have

$$\frac{\Delta U_i}{\Delta z} = \epsilon Z \frac{0.65}{\pi^2} E_{ind} \frac{J_1}{J_i} \frac{a}{d} M_e M_i . \qquad (2)$$

The first harmonic of electron beam current is

$$J_1 = M_1 J_e ,$$

where J_e is the average electron beam current. M_1 is the first harmonic factor.

$$M_1 \simeq 2\frac{\sin\alpha}{\alpha} ,$$

and α is the phase length of bunches. As a rule, $M_1 \simeq 1.5-2$ for powerful klystrons.

The Ion Intensity

Maximal accelerating gradient of the two-beam accelerator is defined by shunt impedance of unloaded cavities and is achieved at a small value of the ion current. At a high intensity of the ion beam the accelerating gradient becomes to be proportional to the ratio J_e/J_i. For two-beam accelerator with a high efficiency of conversion the electron beam energy into the ion beam energy, η_{ei}, the average ion beam current is equal to

$$J_i = 0.65 \frac{\eta_{ei}}{\sqrt{1-\eta_{ei}}} \sqrt{\frac{E_{ind} J_1 M_e a^2}{2\pi\Theta_i Q_0 \rho_0 d}} , \qquad (3)$$

where $\rho_0 = \sqrt{\mu_0/\varepsilon_0} = 120\pi$. Equation (3) describes correctly initial stage of acceleration when $L = \beta\lambda$. For the medial and final parts a compression of the ion and electron beams may be used and the limitation of the ion beam current is important for the initial part only.

The Compression of the Ion and Electron Beams

At a higher energy of the ions, when condition $\beta_e = \beta_i$ may be realized in practice, a possibility of increasing the average current of ion and electron beams is achieved by means of compression of a bunch train in time. This fact matches well with an opportunity of the pulsed power compression in the induction linac. The beam compression is realized by means of a creation a tilt of the ion and the electron velocities. To achieve this purpose it is necessary that:

1. The average current of the electron beam is increased in time. In this case the induced rf voltage in the cavity (and the accelerating rate of ions) will be increased in time.

2. The accelerating voltage of the induction linac is increased in time to provide necessary tilt of the electron velocity.

The present day technique of LIA allows one to compress the pulses down to 10 – 20 ns. However, the increase of the average electron current in the bunch train may limit the level of the compression in LIA. But the compression process may by continued after LIA because of the ion velocity tilt.

The Ion Focusing

Longitudinal magnetic field can be used to focus the electron and ion beams in the medial and final part of the two-beam accelerator. One possible way of the ion focusing at the initial stage of acceleration may be to use a space charge of the electron beam. When $\beta_i < \beta_e$ electron bunches outstrip the ions (at $\beta_e = -\beta_i$ meet the ions), that focusing forces of the space charge affect on ion beam. The repetition rate of the focussing pulses is much higher than the frequency of transverse oscillations of ions and the ion focusing is determined by the average current of the electron beam J_e. The ion oscillation frequency in the potential well of the electron beam with current J_e and radius r_e is equal to

$$\omega_i = \sqrt{\frac{1}{2\pi} \frac{Z}{A} \frac{\epsilon \rho_0 J_e}{Mc^2} \frac{c}{r_e}} \ .$$

Equilibrium radius of the ion beam with emittance ϵ_i is equal to

$$r_i^2 = \frac{\epsilon_i \beta_i r_e}{\pi} \sqrt{2\pi \frac{A}{Z} \frac{Mc^2}{\epsilon \rho_0 J_e}} \ .$$

TWO-BEAM LINAC TEST FACILITY

The induction linear accelerators at JINR may be upgraded for two-beam accelerators without significant expenses. Upgrading the linear accelerator of electron-ion

TABLE 2. Design Parameters of Two-Beam Accelerator Test Facility

eU_{ion} MeV	$E_i nd$ kV/m	Q_0	τ μs	J_{ion} mA	E_{ion} MV/m
	10		6.0	79	8.1
2.0	50	4200	3.0	180	18
	100		0.6	250	26
	10		6.0	64	10
5.0	50	6800	3.0	140	22
	100		0.6	230	32
	10		6.0	48	13
20	50	11000	3.0	110	30
	100		0.6	150	42
	10		6.0	40	13
50	50	16000	3.0	90	30
	100		0.6	130	42
	10		6.0	36	18
100	50	20000	3.0	80	40
	100		0.6	110	56

rings LUEK-20 is easier than others LIA at JINR. This wide aperture induction accelerator has ceramic accelerating tubes without diaphragm. The inner diameter of the ceramic tubes is equal to 130 mm. To upgrade LUEK-20 for two beam accelerator, the electrodynamic structures have to be installed inside smooth ceramic tubes. These structures contain the row of the cavities and rf absorbing insertions. The insertions distribute the induction voltage and suppress parasitic oscillations of the rf structure [9,10]. In paper [18] the application of the LUEK-20 module for an acceleration of a high current electron beam is reported. At 1.5 kA beam current the accelerating gradient is equal to 0.8 MV/m and the pulse duration is 60 ns. Using these parameters of the accelerator, we calculated expected parameters of the two-beam accelerator which may be constructed on the base of existent LUEK-20 accelerator (see Table 2). Parameters of the Test Facility have been calculated at the following data: radius of the cavity $a = 6.1$ cm, diameter of drift tubes $d = 3$ cm, rf resonance wavelength 16 cm, driving electron beam current J_e =500 A, amplitude of the first harmonic of the current $J_1 = 1.5 J_e$, $M_e = 1$ and radius of the electron beam $r_e = 2$ cm. Parameters of the ion beam are $\Theta = \pi$, $M_i = 0.64$ and $\epsilon = 100$ mm mrad.

ACKNOWLEDGEMENTS

The author is grateful to A.A. Fateev, I.N. Ivanov and M.V. Yurkov for many useful discussions.

REFERENCES

1. Dolbilov G.V. et al.,"Experiments on Acceleration of Nitrogen Ions in a Prototype of JINR Heavy Ion Collective Accelerator", *Proc. of the III Int. Symp. on Collective Methods of Acceleration, Lagune Beach California, 1978*, **vol. 2**, p.83-89.
2. Gorinov B.G. et al.,"Experimental Study of the Systems of Induction Accelerator of Enhanced Cyclicity,SILUND-2", *Preprint JINR-9-12148*, Dubna (1979).
3. Dolbilov G.V. et al."Acceleration of Electron Ion Ring in Electrical Field",*Preprint JINR-P9-12414*,Dubna (1979).
4. Dolbilov G.V. et al.,"Head Model of SILUND-20 Linear Induction Accelerator",*Preprint JINR-9-82-339* (1982).
5. Aleksandrov V.S. et al.,"The First Stage of Heavy Ions Collective Accelerator KUTI-20",*Proc.of XIII Int. Conf. on High Energy Accelerators, Novosibirsk, 1986*, Nauka, **vol.1**, p.241-43, (1987).
6. Dolbilov G.V. et al., "SILUND-20 Electron Linear Induction Accelerator", *Preprint JINR-P9-86-290* (1986).
7. Fateev A.A., "Powerful Nanosecond Modulators for Linear Induction Accelerators at JINR", *Presented on Workshop on Space Charge Dominated Beam Physics for Heavy Ion Fusion*, RIKEN (1998).
8. Dolbilov G.V. et al.,"The KUTI-20 Accelerator First Stage Adjusting", *Proc. of Linear Accel. Conf.*, Stanford, SLAC, **XXX**, p.620 (1986), (SLAC-303).
9. Dolbilov G.V. et al.,"A Concept of a Wide Aperture Klystron with Absorbing Drift Tubes for a Linear Collider",*Nuclear Instruments and Methods in Physics Research*,**A**, **383**, p.p.318-324 (1996).
10. Dolbilov G.V. et al.,"Achievement of 100 MW Output Power in a Wide Aperture VLEPP Klystron with Distributed Suppression of Parasitic Modes.
11. Gardelle J. et al.,"Experimental Coupling of 35 GHz RF-Cavity with an Intense Bunched Electron Beam",*Proc. of 6th European Particle Accelerator Conference*, Stockholm (1998)
12. Kaminsky A.A. et al.,"Development of Millimeter-Wave FEM for electron-positron colliders",*Proc. of 6th European Particle Accelerator Conference*, Stockholm (1998)
13. The NLC Design Group,"An RF Power Source Upgrade to the NLC Based on the Relativistic-Klystron", *In Zero-Order Design Report for the Next Linear Collider*,LBNL-5424, SLAC-474, UCRL-ID-124161, UC-414 (19960)
14. Elzhov A.E. et al., "New Scheme of Two Beam Accelerator Driver Based on Linear Induction Accelerator", *Proc. of 6th European Accel. Conf.*, Stockholm (1998)
15. Lebedev I.V. *UHF Technics and Device*, Moscow: "Vysshaya Skola", v.2, 1972.
16. Meinke H. and Gundlach F.W. *Taschenbuch der Hochfrequenztechnic*, Moscow-Leningrad, v.1, 1961
17. Kapchinsky I.M. *Beam Dynamics in Linear Resonance Accelerators*, Moscow, "Atomizdat", 1966
18. Abubakirov E.B. et al.,"Generation and Acceleration of High Current Hollow Electron Beam in Linear Induction Accelerator and Generation of Powerful Microwave Radiation from Cherenkov TWT", *Proc. of II European Conference on Charge Particle Accelerators*, **v.I**, p.34, Nice (1990)

Powerful Nanosecond Pulsed Generators for Linear Induction Accelerators at JINR

G.V. Dolbilov, A.A. Fateev, V.A. Petrov, A.I. Sidorov

Joint Institute for Nuclear Research, 141980 Dubna, Moscow Region, Russia

Abstract.
The paper presents a review of nanosecond pulse generator schemes for LIA developed at the JINR. The main feature of these schemes consists in the use of relatively low-voltage generators ($V \sim 20 - 50$ kV) with low-resistance output impedance ($R \sim 0.5$ Ω). A high power in nanosecond pulses ($W \sim 1$ GW) is produced by nonlinear compression schemes with distributed parameters which compress electromagnetic energy in time.

INTRODUCTION

More than thirty years ago a program of collective methods of acceleration has been started at the Joint Institute for Nuclear Research (JINR). The idea of the method was to accelerate heavy ions in the collective field of intensive electron bunch produced by a linear induction accelerator (LIA) [1]. Several linear induction accelerators have been constructed in the framework of this program [2-7]. This report presents an overview of powerful modulators developed for LIA at JINR. Several different schemes of modulators have been realized and tested experimentally (see Table 1).

TABLE 1. Modulators of the JINR Induction Accelerators

Name of Accelerator	Date of design	P GW	U kV	τ ns	ρ Ω	f Hz	ref.
SILUND	1971-73	0.1	20	20	3.5	1	[1]
SILUND-10	1978-80	2.2	40	20	0.7	1	[9]
SILUND-20	1981-82	0.6	17	30	0.5	50	[4]
LUEK-20	1985-86	5	50	80	0.5	20	[5]

P – output power, U – output voltage, ρ – output impedance, τ – pulse duration, f – repetition rate

CP480, *Space Charge Dominated Beam Physics for Heavy Ion Fusion*, edited by Yuri K. Batygin
© 1999 The American Institute of Physics 1-56396-860-6/99/$15.00

Modulator for the linear induction accelerator should meet several technical requirements. The main of them are requirements for a high peak power, high repetition rate and reliability. The problem of a high repetition rate has been solved by means of using hydrogen thyratrons as commutating elements. An effective approach to solve the problem of a high peak power has been proposed and successfully realized for the first time at JINR [3]. The idea consists in using nonlinear magnetic elements as power compressors. At present this approach is used widely in different organizations. The problem of reliability has been solved by means of decreasing the output voltage of the modulators. An unwilling consequence of such an approach is a low output impedance, which leads to a requirement of fine tuning of all modulator elements, especially those of output stages.

SILUND

The accelerator SILUND (commissioned in 1973) was the first one designed especially for collective method of acceleration (see Fig. 1). It accelerated intense electron beams with short pulse duration, $\tau \sim 10$ ns, at the repetition rate of 1 pps. The modulator of this accelerator provided driving pulses with short rise time and stability of the voltage of about 2 – 3 %.

The simplified scheme of the modulator is presented in Fig. 2. Main elements of the scheme are the energy storage capacitor C, the thyratron commutator T and three sharping nonlinear ferromagnetic lines NL followed by inductors with ferrite cores. The initial pulse with the rise time of about 100 ns is applied to the inputs of the lines filled with ferrite and liquid dielectric. The shock electromagnetic wave with rise time of about 3 ns is produced in each line by means of energy dissipation

FIGURE 1. The layout of the accelerator SILUND.

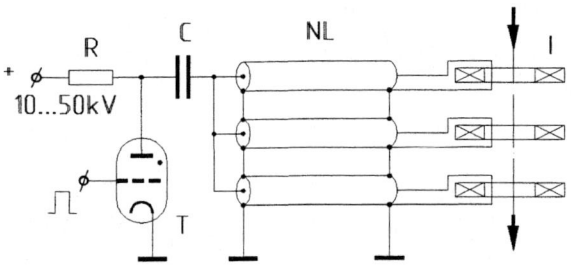

FIGURE 2. Scheme of the SILUND modulator. Here NL is nonlinear ferromagnetic lines and T is hydrogen thyratron.

at the pulse front. The output pulse has almost rectangular form with duration determined by the saturation of the inductor core. The total pulse power at the inductor inputs can exceed the power commutated by the thyratron at certain conditions, but the amplification ratio is not larger than 2 [8]. The scheme is simple and reliable but provides relatively low efficiency and pulse power limited by thyratron resources.

SILUND-20

To increase the efficiency and pulse power of the modulators, it has been decided to use power compression schemes with nonlinear ferromagnetic elements. Such schemes use the large changes of permeability of ferromagnetic materials of the nonlinear reactors. The initial pulse propagates through several compression stages and becomes shorter in time and higher in power. The compression ratio is approximately proportional to the volume of ferromagnetic material and inversely proportional to the square root of the pulse energy. Thus, pulse compressors are mostly effective in the nanosecond range when total pulse energy is relatively low. But it is difficult to form rectangular nanosecond pulses without considerable energy losses.

General feature of such nonlinear schemes is that there exists significant delay between the commutator switching and output pulse. This delay substantially depends on the charge voltage and initial magnetization of ferromagnetic. As a result, the problem to achieve required level of synchronization of output pulse in nanosecond range becomes to be complicated. Nevertheless, our experience has shown that this problem can be solved by means of an precise tuning of the initial magnetization of the nonlinear cores from cycle to cycle and of the charge voltage.

The SILUND-20 accelerator was the first one where the compression technique has been used in the modulator design [3–6].

The scheme of the SILUND-20 modulator is presented in Fig. 3. It operates as follows. The initial pulse is compressed in a conventional stage. Then this pulse

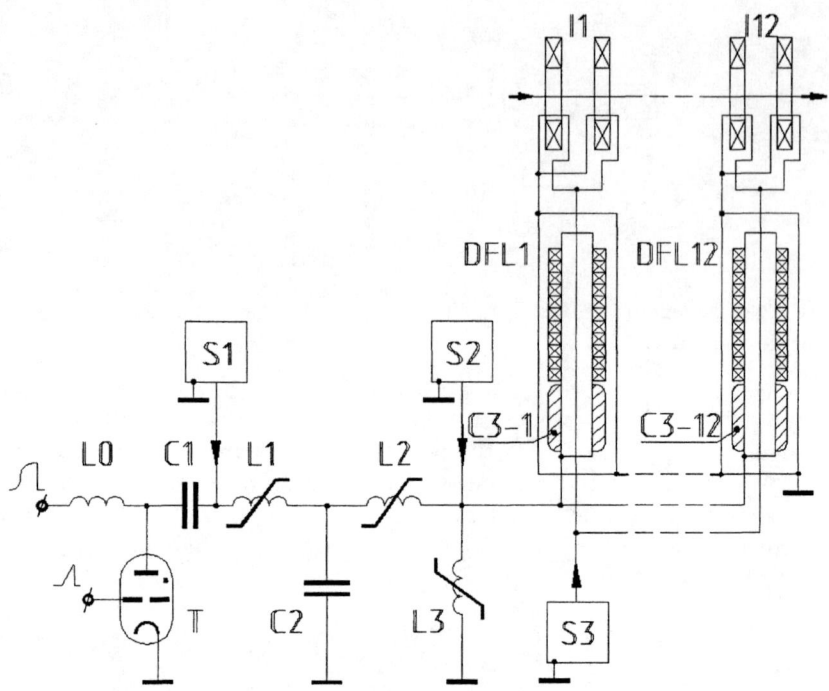

FIGURE 3. Scheme of the SILUND-20 modulator. Here T is hydrogen thyratron, L1 – L3 are nonlinear reactors, DFL1 – DFL12 are nonlinear double forming lines, I1 – I12 are inductors and S1 – S3 are ambipolar pulse current sources.

charges the forming stage consisting of nonlinear double forming lines (DFL1 – DFL12) with additional correcting capacities C3-1 – C3-12. Reactor L3 is not saturated and the transmitted current is considerably less than the charge current during this process. Double forming line, DFL, consists of two coaxial lines. The outer line of the DFL is partially filled with ferrite rings. The rest volume of the DFL is filled with glycerine. Ferrite is saturated with respect to the charge current. The efficiency of the energy transfer is of about 90 %.

When the voltage amplitude in the DFL reaches its maximum, the reactor L3 is saturated and becomes to serve as a commutator forming a voltage swing with the relatively long front which then becomes shorter and forms $\sim$ 5 ns rise time of the driving pulse.

The initial magnetization of the ferromagnetic materials in the nonlinear elements is adjusted by three ambipolar current sources (S1 – S3). This technique provides a high precision control of the initial magnetization value. Together with the high precision ($\sim$ 0.04 %) of the initial voltage supply, this dejittering technique provides the output pulse time instability less than 1 ns.

FIGURE 4. The layout of the SILUND-20 modulator.

The layout of the modulator is presented in Fig. 4 and its parameters are presented in Table 1.

LUEK-20

The modulator with a longer output pulse duration has been constructed for the LUEK-20 accelerator [7].

The scheme of the modulator is shown in Fig. 5. Hydrogen thyratron T with a grounded grid has been used as a switch in the initial stage (1). It allows to increase

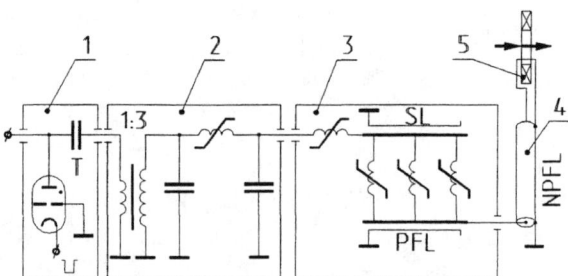

FIGURE 5. Scheme of the LUEK-20 modulator. Here: 1 – initial stage, 2 – compression stage with lumped elements located at oil filled volume, 3 – compression stage located at water filled volume, SL and PFL – strip lines, 4 – ferromagnetic lines, 5 – permalloy inductors.

FIGURE 6. The layout of the LUEK-20 modulator.

the discharge current value by several times without substantial reduction of the operation resource. The second stage (2) is designed as a traditional compression circuit immersed in oil filled volume.

The next stage consists of 6 identical blocks (3) connected in parallel. Rectified water is used as a dielectric in the intermediate charge storage SL (storage line) and in the pulse forming line PFL. The storage line SL and the pulse forming line PFL are fabricated as strip lines.

Each block is connected by 9 (only one is shown in Fig. 5) nonlinear pulse forming lines (4) with the block of permalloy inductors (5). If the nonlinear lines are saturated, their total wave resistance is equal to that of the PFL. The modulator feeds one accelerating section consisting of 36 inductors. The total output wave resistance of the modulator is equal to 0.5 Ω. The maximum pulse power provided by the modulator is equal to 5 GW. The output voltage is up to 50 kV. Pulse duration is up to 80 nanoseconds. The layout of the LUEK-20 modulator is presented in Fig. 6.

SILUND-10

Some novel ideas on powerful nanosecond modulators have been realized during the design and construction of the high current accelerator SILUND-10. To increase the intensity of accelerating beams and the peak power it is necessary to increase considerably the volume of ferromagnetic material per the unit of the accelerator length. Here the stray inductances of all the elements and couplings have to be

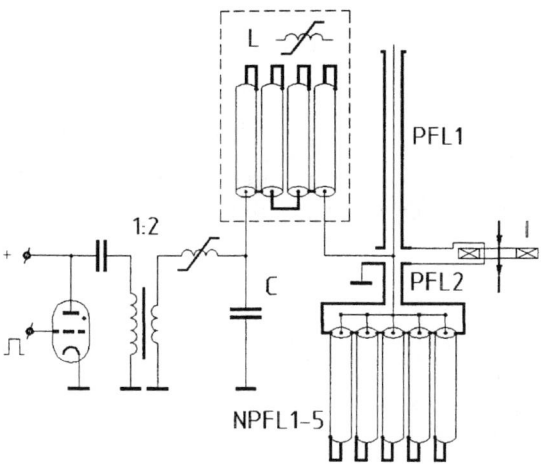

FIGURE 7. Scheme of the SILUND-10 modulator. Here L – nonlinear reactor consisting of the short-grounded ferromagnetic lines connected in series, NPFL1-5 – commutating block consisting of the short-grounded ferromagnetic lines connected in parallel, PFL1 and PFL2 – linear forming lines.

FIGURE 8. The layout of the SILUND-10 modulator.

minimized.

These problems have been solved during the design and construction of the SILUND-10 accelerator by means of the short-grounded nonlinear coaxial ferromagnetic lines which are used as switches (see Fig. 7). The lines are connected seriesly in the prior stage (L in Fig. 7) and in parallel – in the forming stage.

The forming stage consists of DFL (pulse forming lines 1 and 2) switched by the short-grounded nonlinear pulse forming lines (NPFL1-5). If nonlinear lines are saturated, the both parts of the DFL become to be matched as in the wave resistance as in the electrical length.

Experimental setup consists of one modulator and the induction accelerating module. Each of six inductors of the accelerating module is driven by two DFL. The total accelerating voltage $U = 250$ kV has been obtained at the 8 kA equivalent load. The layout of the high current accelerating modulus is presented in Fig. 8.

CONCLUSION

The development of the high power modulators at JINR has been performed in the framework of general concept. Hydrogen thyratrons with high stability of their parameters in time, reliability and precision of synchronization $\sim$ 1 ns, are used as a commutators. The traditional compression stages with lumped elements are used in the primary stages of power compression. Short voltage swings are produced by means of nonlinear ferrite lines. The voltage of the driving pulses is relatively low and usually is of about 20 – 50 kV. As a result, the output impedance of the modulator is relatively low. This is achieved by using a large number of final circuits connecsed in parallel. This concept of nanosecond modulators for LIA has proved to be fruitful and has been realized in modulators of the SILUND, SILUND-20, SILUND-10 and LUEK-20 accelerators.

REFERENCES

1. Dolbilov G.V. et al.,"Experiments on Acceleration of Nitrogen Ions in a Prototype of JINR Heavy Ion Collective Accelerator", *Proc. of the III Int. Symp. on Collective Methods of Acceleration, Lagune Beach California, 1978,* **vol. 2**, p.83-89.
2. Beznoshchenko N.I. et al., "High Current Induction Linear Accelerator", Proc. of the Fourth All-Union Conf. on Charged Particle Accelerators (Moscow, 18-20 November, 1974), vol.I, Moscow, "Nauka", 1975, p.290.
3. Gorinov B.G. et al.,"Experimental Study of the Systems of Induction Accelerator of Enhanced Cyclicity, SILUND-2", *Preprint JINR-9-12148*, Dubna (1979).
4. Dolbilov G.V. et al.,"Head Model of SILUND-20 Linear Induction Accelerator",*Preprint JINR-9-82-339* (1982).
5. Aleksandrov V.S. et al.,"The First Stage of Heavy Ions Collective Accelerator KUTI-20",*Proc.of XIII Int. Conf. on High Energy Accelerators, Novosibirsk, 1986*, Nauka, **vol.1**, p.241-43, (1987).

6. Dolbilov G.V. et al., "SILUND-20 Electron Linear Induction Accelerator", *Preprint JINR-P9-86-290* (1986).
7. Dolbilov G.V. et al.,"The KUTI-20 Accelerator First Stage Adjusting", *Proc. of Linear Accel. Conf.*, Stanford, SLAC, **XXX**, p.620 (1986), (SLAC-303).
8. Fateev A.A., "The Pulse Power Amplification in Nonlinear Ferromagnetic Lines Sov. J. Pribory i Tehnika Eksperimenta, N1, 1988, p.101.
9. Dolbilov G.V. et al.,"Development of Nanosecond Pulsed Generators for Linear Induction Accelerators at JINR", *Proc. of the 1994 International Linac Conf.*, Tsukuba, Japan, 1994, v.1, p.360-62.

Inhomogeneity Smoothing Using Density Valley Formed by Ion Beam Deposition in ICF Fuel Pellet

Kazuhiro Fujita, Takashi Kikuchi, Daisuke Takahashi,
Masaru Yazawa and Shigeo Kawata

*Department of Electrical Engineering, Nagaoka University of Technology,
Nagaoka 940-2188, Japan*

Abstract.
We study the beam non-uniformity smoothing effect of the radiation transport in the density valley formed by an ion-beam deposition in an ion-beam inertial confinement fusion pellets by numerical simulation. The simulation results show that the radiation energy is confined in the density valley, and the beam non-uniformity can be smoothed out by the radiation transport along the density valley. The formation of density valley is controlled by changing a beam incident angle. Consequently, the simulation results show that the radiation smoothing can be also controlled by the density valley structure.

INTRODUCTION

In inertial confinement fusion(ICF), one of most important issues is to find a way to implode a fuel pellet in a spherically symmetrical manner [1–6]. In this paper, we studied a radiation-smoothing effect on an implosion symmetry under a non-uniform acceleration for a direct-driven pellet. In an ion beam pellet implosion, ion beams may introduce an implosion non-uniformity. The non-uniformity introduces the non-uniform implosion and degrades the fusion energy output [7–9]. Even in a direct-driven reactor-size pellet, the radiation smoothing is important. In addition, an ion deposition layer is rather thick in ion-beam ICF. It may be several hundred μm, although it depends on particle energy and material. A pressure peak appears in a rather deep part of a deposition layer. Therefore, a density valley is formed in the energy absorber. This density valley has a role of the radiation confinement and we can also expect the smoothing of the beam non-uniformity in this valley.

First, we show a simulation model employed in this paper. Second, the radiation non-uniformity smoothing effect is studied by the computer simulation. It confirms that a density valley is formed by the ion beam deposition and the initial non-uniformity can be smoothed by the radiation transport in the density valley. Third, the simulation results also shows that the formation of the density valley can be controlled by changing the beam incident angle. Consequently the radiation non-uniformity smoothing effect can be controlled by the ion-beam deposition.

SIMULATION MODEL

In order to study the non-uniformity smoothing effect of the radiation transport on an ion-beam pellet implosion, we employed a two-dimensional(r, θ) Lagrangian hydrodynamic code. The physical model employed in this paper is based on a two-temperature(plasma and radiation temperatures) fluid model.

An ion-beam ICF pellet employed in this paper is presented in Fig. 1, and consisted of three layers of Pb, Al, and DT. The DT fuel contained is 5mg, and the total masses of D and T are equal. The outer radius of the pellet is 5mm. The initial state of Pb, Al and DT are in solid. An ion(proton)-beam energy driver parameter in this paper is presented in Fig. 2. The total ion-beam input energy is 7MJ. The ion-beam pulse duration is 40ns. The beam power increases with a function of $(time)^{2.5}$. The ion-beam deposition energy is computed by solving the equations for ion stopping power including the plasma effect. In this paper, a beam ion impinges the pellet surface with an incident angle, and the incident angle distribution is uniform between the normal and the maximum incident angle.

The non-uniformity is introduced by changing the ion-beam number density as follows:

$$n = n_0(1.0 + \delta[\cos(m\theta)]) \qquad (1)$$

where δ is the amplitude of the non-uniformity and m is the mode number. In this paper, the mode number is 2 in the simulation.

NON-UNIFORMITY SMOOTHING EFFECT OF RADIATION TRANSPORT

Figure 3 presents the density profile in space. The beam non-uniformity is 2.5% and the maximum incident angle is 45 degree. The beam non-uniformity introduces the implosion non-uniformity. The low-density part exists in the Al layer. This is called the density valley in this paper and the beam non-uniformity is smoothed by the radiation transport along the density valley.

Figure 4(a) presents the density and the radiation temperature profiles as a function of radius at 30 nsec; a solid line shows the density profiles and a dotted line shows the radiation temperature. It is shown that the density valley is formed and the radiation is confined in the valley. Figure 4(b) presents the peak radiation temperature profiles in the Al layer at 45nsec as a function of θ; a solid line shows a result in which a radiation transport is included normally and a dotted line shows one including a radiation transport only in the radius direction. The beam non-uniformity can be smoothed out by the radiation transport from 2.5% to 1.6%.

Figure 5 presents the fusion output energy as a function of the beam non-uniformity; a solid line shows a result in which a radiation transport is included normally and a dotted line shows one including a radiation transport only in the radius direction. The beam non-uniformity can be smoothed well, and that the maximum tolerable beam non-unformity is 2.5% in this study. Table 1 also shows the simulation results, the fusion output energy, ρR, the maximum ion temperature and the implosion efficiency, which support the result shown in Fig. 5.

These simulation results for a pellet implosion demonstrate that the radiation transport may smooth out the non-uniformity in the density valley even in the direct-driven ICF pellet.

CONTROL OF DENSITY VALLEY FORMATION

In the previous section, it was found out that the beam non-uniformity can be smoothed in the density valley. The simulation study also shows that the density valley structure is controlled by changing the beam maximum incident angle; the radiation confinement in the valley and the radiation transport along the valley are also controlled by changing the valley structure.

Figure 6 presents the density profiles as a function of radius at about 20 nsec, 30 nsec and 40 nsec. The density valley is deep and narrow, when the maximum beam incident angle is small. The density valley is flat and wide, when the maximum beam incident angle becomes large. Therefore the density valley structure can be controlled by changing the beam incident angle. Table 2 shows that the fusion energy output, ρR and the maximum ion temperature become large, when the maximum incident angle becomes large. Although the implosion efficiency becomes small, when the maximum incident angle becomes large. Figure 7 presents the radiation temperature profiles as a function of radius at about 20 nsec, 30 nsec and 40 nsec. In Fig. 7, the radiation temperature becomes large, when the maximum incident angle becomes large. Therefore the radiation is confined better in the density valley when the maximum incident angle becomes large. In addition, Figure 8 presents the thermal conductivity profiles as a function of radius at about 30 nsec and 40 nsec. The thermal conductivity is certainly high, when the maximum incident angle becomes large. Therefore the radiation transport can be more smoothed along the density valley, when the maximum incident angle becomes large. Consequently, the beam non-uniformity can be more smoothed out along the density valley by the radiation transport, when the maximum incident angle becomes large. Table 3 also shows the simulation results, which support the results mentioned above. In Table 3, the radiation energy confined in the valley and the thermal conductivity along the valley become large, when the maximum incident angle becomes large. Therefore the effective non-uniformities becomes small when the maximum incident angle increases.

These simulation results for a pellet implosion demonstrate that the density valley structure can be controlled by changing the beam incident angle. The radiation confinement in the density valley and the radiation transport along the density valley are also controlled by changing the valley structure. The beam non-uniformity can be more smoothed out along the density valley by the radiation transport when the maximum incident angle becomes large.

CONCLUSIONS

In this paper, we studied the radiation smoothing effects on a pellet implosion by computer simulations. In addition, we studied also that the density valley structure is controlled by changing the beam maximum incident angle. The first result obtained that the beam non-uniformity can be smoothed out by the radiation transport along the density valley. In a direct-driven ion-beam ICF pellet, the temperature of the ablation part is about several hundred eV and the density is rather high. In the density valley in the energy absorber, the radiation is confined and plays a roll to smooth out the beam non-uniformity. We also found that the density valley structure is controlled by changing the beam maximum incident angle; the radiation confinement in the density valley and the radiation transport along the density valley are also controlled by changing the valley structure. The simulation results present in this paper that the beam non-uniformity can be more smoothed out along the density valley by the radiation transport, when the maximum incident angle becomes large.

REFERENCES

1. Emery, M.H. et al., *Phys. Rev. Lett.* **48**, 253 (1982).
2. Kawata, S., and Niu, K., *J. Phys. Soc. Jpn.* **53**, 3416 (1984).
3. Mark, J.W.-K., *Proc. 7th Int. Conf. on High-Power Particle Beams*, 1988, vol. 1, p.785.
4. Tabak, M., and Callahan-Miller, D., *Phys. Plasmas* **5**, 1895 (1998).
5. Tabak, M., Callahan-Miller, D., Ho, D. D.-M., and Zimmerman, G. B., *Nucl. Fusion* **38**, 509 (1998).
6. Kawata, S., Sato, T., Teramoto, T., Bandoh, E., Masubuchi, Y., and Takahashi, I., *Laser and Particle Beams* **11**, 757 (1993).
7. Tahir, N. A., Long, K. A., Laing, E. W., *J. Appl. Phys.* **60**, 898 (1986).
8. Kawata, S., Nakashima, H., *Laser and Particle Beams* **10**, 479 (1992).
9. Kawata, S., *Laser and Particle Beams* **13**, 383 (1995).

TABLE 1. The non-uniformity smoothing effect of the radiation transport. When the radiation transport is included normally, the beam non-uniformity can be smoothed well, and the maximum tolerable beam non-unformity is 2.5%.

Radiation	Nonunifomity [%]	OutPut Ene. [MJ]	ρR [g/cm2]	Ti max [kev]	Imp.Eff. [%]
ON	0.0	829	2.42	40.2	2.51
	1.0	831	2.37	34.8	2.47
	2.0	820	2.42	32.9	2.50
	2.5	748	2.35	37.0	2.48
OFF	1.0	815	2.37	33.7	2.48
	2.0	751	2.47	37.7	2.47
	2.5	10.2	2.07	2.38	2.44

TABLE 2. Simulation results by changing the beam maximum incident angle. When the maximum beam incident angle becomes large, the fusion energy output, ρR and the maximum ion temperature becomes large.

Max. Incident Angle [deg]	OutPut Energy[MJ]	ρR [g/cm^2]	Ti max [kev]	Imp.Eff. [%]	void closing time [nsec]
45	748	2.35	37.0	2.47	47.5
30	410	1.08	27.4	2.44	47.0
22.5	145	1.01	13.5	2.49	46.7
20	11.1	1.01	2.44	2.47	46.7
15	7.44	0.963	2.06	2.59	46.5
0	4.60	0.940	1.69	2.56	46.4

TABLE 3. The effective normalized non-unformity ($\Delta \rho$, Δp and Δu), the radiation energy and the thermal conductivity by changing the beam incident angle. The results are normalized by the values at 45 degree and are averaged in space and time.

Max. Incident Angle [deg]	$\Delta \rho$ in DT	Δp in DT	Δu in DT	Radiation Energy	Radiation Thermal Conductivity
45	1.00	1.00	1.00	1.00	1.00
30	1.03	1.12	1.03	0.929	0.975
22.5	1.07	1.08	1.07	0.902	0.961
20	1.17	1.18	1.09	0.891	0.955
15	1.28	1.18	1.05	0.864	0.899
0	1.49	1.48	1.21	0.842	0.864

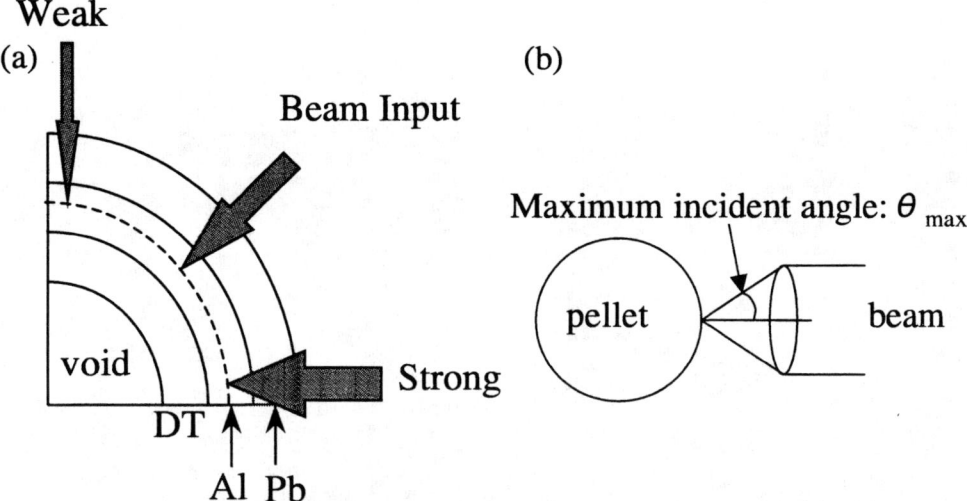

FIGURE 1. (a)The fuel pellet structure employed in the simulation in this paper. The ion-beam impinges the fuel pellet and deposits its energy mainly in Al. The Pb layer behaves as a tamper and the Al layer as a beam energy absorber. The inner part of Al behaves as a pusher. (b)An ion-beam has the incident angle θ, and the distribution of beam incident angle is uniform between the normal (0 degree) and the maximum incident angle θ_{max}.

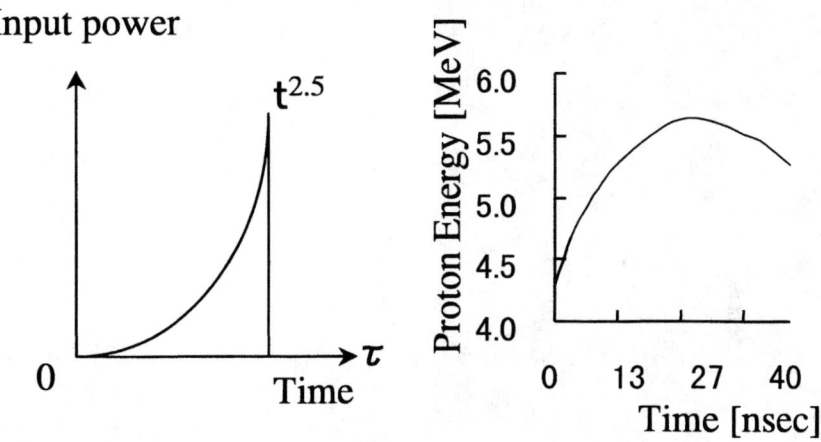

FIGURE 2. An ion-beam energy driver parameter. The total ion-beam input energy is 7MJ. The ion-beam pulse duration is 40ns. The beam power increases with a function of (time)$^{2.5}$. The particle energy is controlled so that the particles impinging normally to the pellet surface stop at the center of Al layer.

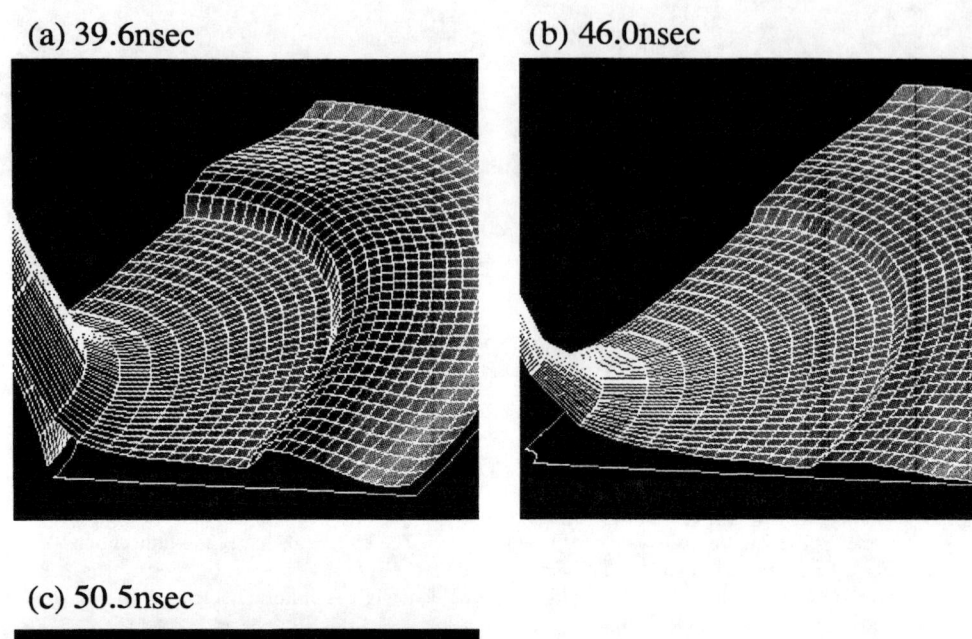

FIGURE 3. Density profiles in space at the 2.5% beam non-uniformity, and the maximum beam incident angle is 45 degree. The beam non-uniformity introduces the non-uniform implosion.

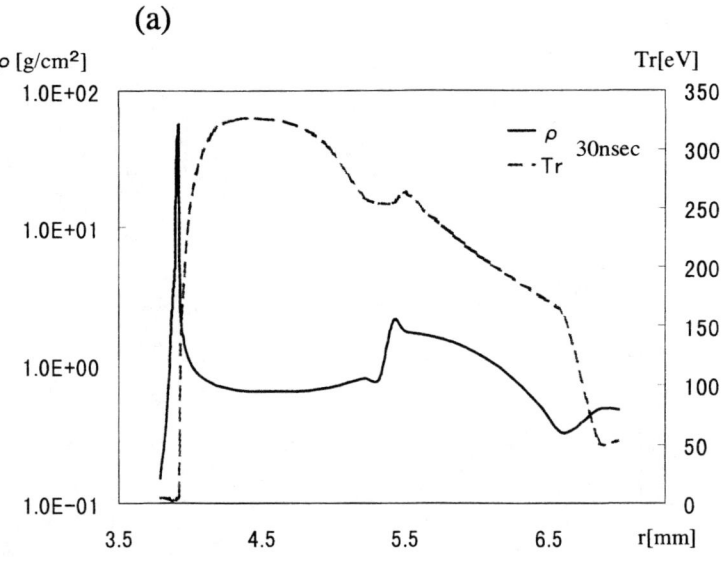

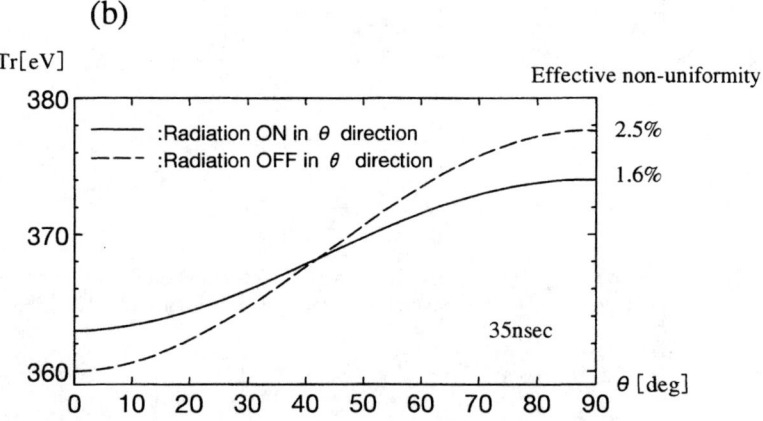

FIGURE 4. (a)The density and the radiation temperature profiles as a function of radius at 30 nsec. The density valley is formed and the radiation is confined in the valley. (b)The peak radiation temperature profiles as a function of θ in the Al layer at 35nsec. The beam non-uniformity can be smoothed out by the radiation transport from 2.5% to 1.6%.

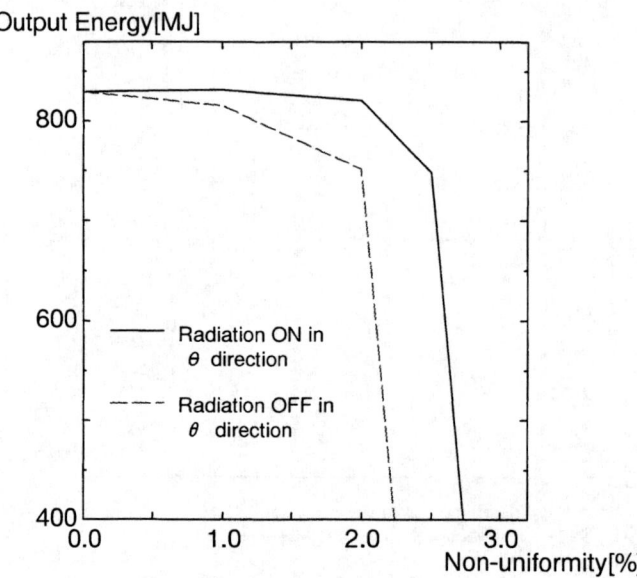

FIGURE 5. The fusion output energy as a function of the beam non-uniformity. The beam non-uniformity can be smoothed well, and the maximum tolerable beam non-unformity is 2.5%.

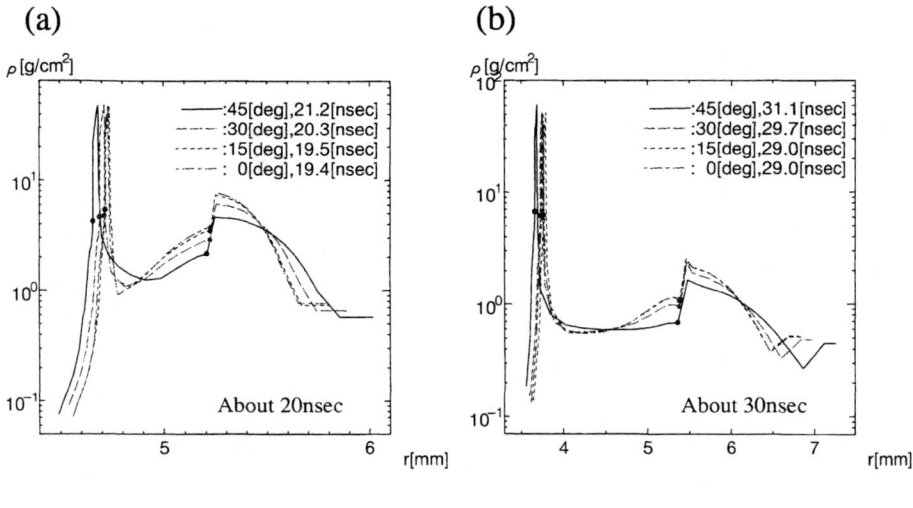
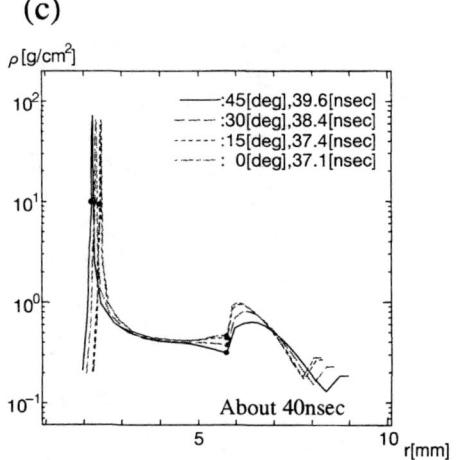

FIGURE 6. The density profiles as a function of radius at about 20 nsec, 30 nsec and 40 nsec. The density valley is deep and narrow, when the maximum beam incident angle is small. The density valley is flat and wide, when the maximum beam incident angle becomes large.

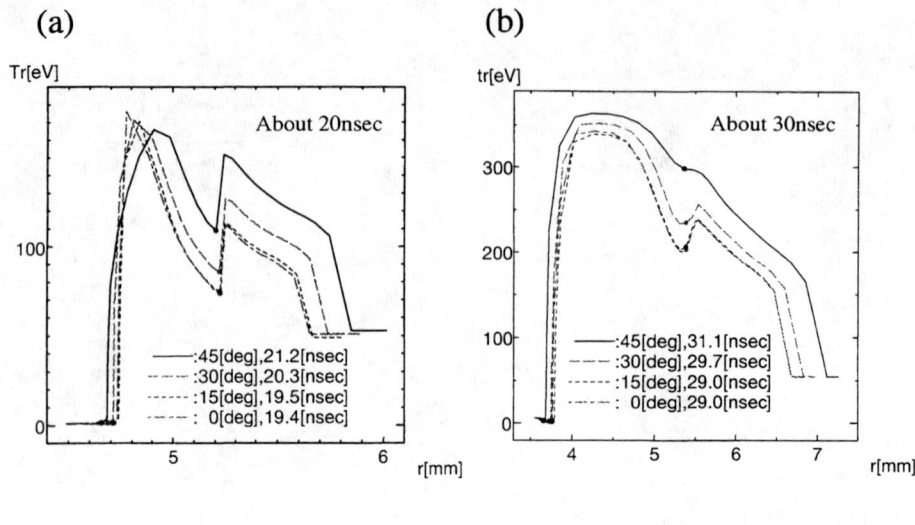
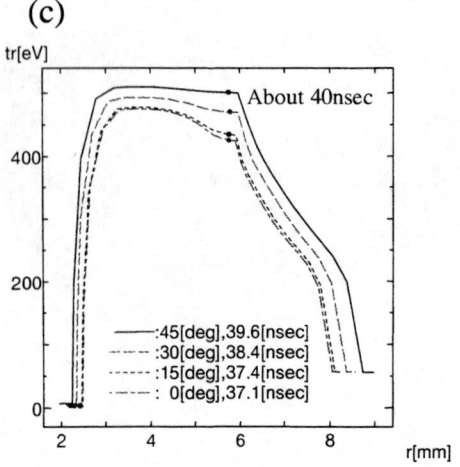

FIGURE 7. The radiation temperature profiles as a function of radius. The radiation temperature becomes large, when the maximum incident angle becomes large.

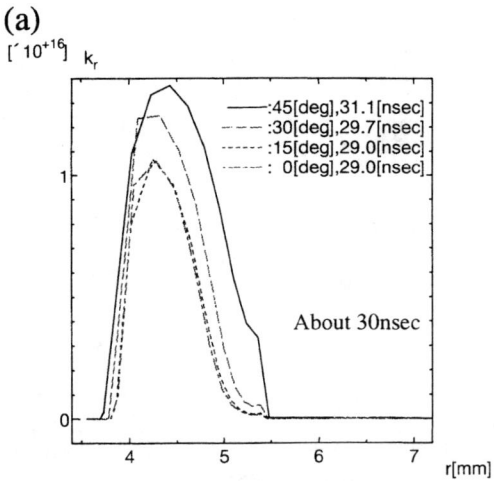
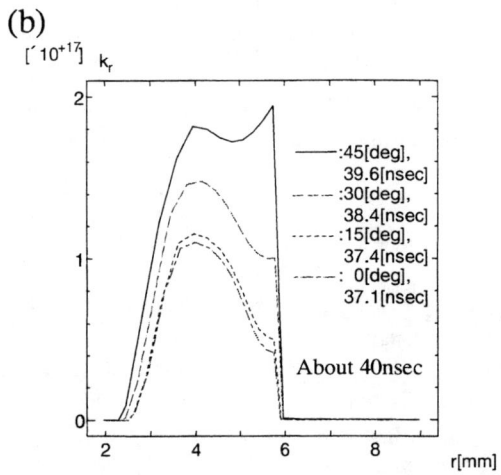

FIGURE 8. The thermal conductivity profiles as a function of radius. The thermal conductivity is certainly high, when the maximum incident angle becomes large. Therefore the radiation transport can be more smoothed along the density valley, when the maximum incident angle becomes large.

Intense-Heavy-Ion-Beam Transport Through an Insulator Beam Guide for Heavy Ion Fusion

Takashi Kikuchi, Shigeo Kawata, Shigeru Kato*,
Susumu Hanamori and Masaru Yazawa

Department of Electrical Engineering, Nagaoka University of Technology, Nagaoka 940-2188, Japan
*Nissin Electric Co., Ltd., Kyoto 615, Japan

Abstract. We study an intense-heavy-ion-beam transport through an insulator beam guide by a numerical simulation in heavy ion fusion. In our previous papers [*Jpn. J. Appl. Phys.* **35**, L120 (1996); *Jpn. J. Appl. Phys.* **37**, L471 (1998)] we proposed a new system for electron and proton beam transport using an insulator beam guide. We apply this system to a singly charged cesium (Cs^+) beam transport and a lead (Pb^+) ion beam transport. An intense heavy ion beam creates a local electric field on the insulator inner surface, and generates a plasma on the surface of the insulator guide. Electrons are extracted from the plasma, and the beam charge is effectively neutralized by the electrons. The electron extraction is self-regulated by the net space charge of the beam. Consequently, an intense heavy ion beam can be efficiently transported by the proposed simple system.

INTRODUCTION

High-voltage and pulsed power charged-particle-beam technologies have been applied to various fields: particle acceleration, nuclear fusion, new material processing, microwave source, X-ray source, etc. Especially a heavy ion beam may be one of promising energy-driver candidates inertial confinement fusion [1,2]. In these fields, physics of particle beam behavior must be clarified. Consequently, we should establish a charged-particle-beam-transport control method for the above purposes [3-8].

In our previous papers [7,8] we proposed a transport system for electron and proton beams through an insulator beam guide. We reported that the particle beam charge is neutralized by electrons or protons extracted from a plasma generated on the surface insulator guide, and the charged-particle-beam is efficiently transported through an insulator beam guide. The particles extracted from the plasma on the surface insulator guide are self-regulated by the net space charge of the beam.

Based on the previous results, we propose to apply the insulator beam guide to an intense-heavy-ion beam transport in this paper. Heavy ion species, for example, Pb and Cs may be candidates in heavy ion inertial confinement fusion [9,10]. Even for other purposes presented above it is useful to find a new transport method for a high-current ($\geq$ kA) heavy ion beam. In this paper we employ a Cs$^+$ ion beam and a Pb$^+$ ion beam in order to demonstrate the viability of the proposed insulator beam guide system. We study the intense heavy ion beam transport by using a numerical particle simulation. The simulation results present that (1) the intense heavy ion beam can be efficiently transported through the insulator beam guide, (2) the beam space charge is neutralized well in a self-regulated manner by the electrons extracted from the plasma generated on the insulator surface, and (3) the proposed system is simple.

PHYSICAL MECHANISM AND SIMULATION MODEL FOR BEAM TRANSPORT USING INSULATOR BEAM GUIDE

The physical mechanism for the beam transport through the insulator beam guide is as follows: the intense heavy ion beam creates a local electric field on the insulator inner surface, the local electric field induces the local discharges, and the plasma is produced on the insulator inner surface. Then electrons are extracted from the plasma generated on the inner surface of the insulator beam guide, because of the beam net charge. The electrons follow the heavy ion beam, and the beam charge is effectively neutralized by the electrons. Therefore, the heavy ion beam propagates efficiently through the insulator beam guide. On the contrary to an intense proton beam, the Cs$^+$ and Pb$^+$ ion beam does not form a clear virtual anode, so that we do not need an initially prefilled plasma gas in the beam transport region [8].

Our simulation model is as follows (see Fig. 1): we assume that the phenomenon concerned is cylindrically symmetric. In this study, we carry out a particle-in-cell (PIC) simulation [11–13]. The PIC code used is a 2.5-dimensional one. The field components (E_r, E_z, B_θ), and the particle position and velocity (r, z, v_r, v_θ, v_z) are solved using the Maxwell equations and the relativistic equation of motion.

The Cs$^+$-ion-beam-parameter values are as follows [10]: the input Cs$^+$ ion beam waveform is shown in Fig. 2(a); the maximum current I_{0max} is 6.25 kA, the particle energy is 4 GeV, the pulse width is 10 ns and the rise and fall times are 2 ns. The initial beam radius r_b is 3 cm. The average longitudinal speed v_{z0} of the beam ions injected is determined by the waveform shown in Fig. 2(b), and the beam temperature is 0.1 eV; v_{z0} is 0.248 c at $z = 0$. Here c is the speed of light in vacuum. At the beam entrance, that is, $z = 0$, the beam ions enter uniformly, and the transport area is in vacuum. The computation area is $0 \leq z \leq Z_l\,(= 2$ m) and $0 \leq r \leq R_l\,(= 5$ cm) (see Fig. 1).

The Pb$^+$-ion-beam-parameter values are as follows [9]: the input Pb$^+$ ion beam waveform is shown in Fig. 2(b); the maximum current I_{0max} is 2.5 kA, the particle energy is 15 GeV, the pulse width is 10 ns and the rise and fall times are 2 ns. The initial beam radius r_b is 3 cm. The average longitudinal speed v_{z0} of the beam ions injected is determined by the waveform shown in Fig. 2(b), and the beam temperature is 0.01 eV; v_{z0} is 0.372 c at $z = 0$. At the beam entrance, that is, $z = 0$, the beam ions enter uniformly, and the transport area is in vacuum. The computation area is $0 \leq z \leq Z_l$ ($= 5$ m) and $0 \leq r \leq R_l$ ($= 5$ cm) (see Fig. 1).

The center line of $r = 0$ is the cylindrical axis, at which the cylindrically-symmetric boundary condition is imposed. The other three outer boundaries are conductors. The relative permittivity ϵ_r of the insulator beam guide is 5. In our simulation, a local plasma is generated on the insulator guide surface, when the magnitude of the electric field exceeds the threshold [5] for the local discharge [7]. The threshold is 1×10^7 V/m in this study.

SIMULATION RESULTS

Firstly we simulate the Cs$^+$ ion beam propagation in a vacuum without the insulator beam guide. The particle maps of the Cs$^+$ beam ions and the beam envelope lines $r(z)$ defined from the following equation are shown in Fig. 3:

$$r = r_b + \frac{I_{0\max} q z^2}{4\pi \epsilon_0 v_{z0}^3 r_b m_i} \tag{1}$$

which is obtained by a simple envelope equation [3]. Here, ϵ_0 is the permittivity of vacuum, q is the charge of the beam ion and m_i is the beam ion mass. In this case, the beam radius tends to expand gradually.

Figures 4 and 5 present the particle maps of the Cs$^+$ beam ions and the electrons emitted from the insulator guide surface. The electrons extracted from the plasma generated on the insulator inner surface move along with the Cs$^+$ ion beam. The electrons effectively neutralize the space charge of the beam, and suppress the radial expansion of the beam. Figure 6 shows the history of the total charges of the beam ions and the electrons in the whole transport region. The total electron charge is self-regulated by the beam net charge. For a simulation without the insulator beam guide, the Cs$^+$ ion distribution in the phase space $(r, P_r/m_0 c)$ is shown in Fig. 7(a). Figure 7(b) shows the Cs$^+$ ion distribution in the phase space $(r, P_r/m_0 c)$ of the beam through the insulator guide. The normalized emittance values for the cases without and with the insulator beam guide are 1.70×10^{-5} and 1.53×10^{-5} at $t = 26.8$ ns, respectively. The initial emittance of the Cs$^+$ ion beam is 1.49×10^{-5} at $t = 10.4$ ns.

For our present result shown in Fig. 3 (without the insulator guide) and Fig. 4 (with the insulator guide), the beam average velocities $<v_j>$ ($j = z, r, \theta$) and the standard deviations σ_j ($j = z, r, \theta$), defined by

$$\sigma_j = \sqrt{\left(\sum_{k=1}^{N}(v_{jk} - <v_j>)^2\right)/N}, (j = z, r, \theta) \tag{2}$$

are Table 1 at $t = 10.4$ ns and Table 2 at $t = 26.8$ ns.

TABLE 1. Cs^+ ion beam average velocities and the standard deviations at $t = 10.4$ ns without and with the insulator guide model.

guide	$<v_z>$	$<v_r>$	$<v_\theta>$	σ_z	σ_r	σ_θ
without	$2.47 \times 10^{-1}c$	$2.68 \times 10^{-4}c$	$8.00 \times 10^{-6}c$	$2.84 \times 10^{-4}c$	$4.71 \times 10^{-4}c$	$3.02 \times 10^{-4}c$
with	$2.48 \times 10^{-1}c$	$1.46 \times 10^{-5}c$	$5.06 \times 10^{-6}c$	$2.68 \times 10^{-4}c$	$4.28 \times 10^{-4}c$	$3.05 \times 10^{-4}c$

TABLE 2. Cs^+ ion beam average velocities and the standard deviations at $t = 26.8$ ns without and with the insulator guide model.

guide	$<v_z>$	$<v_r>$	$<v_\theta>$	σ_z	σ_r	σ_θ
without	$2.47 \times 10^{-1}c$	$9.46 \times 10^{-4}c$	$4.44 \times 10^{-6}c$	$3.39 \times 10^{-4}c$	$7.60 \times 10^{-4}c$	$3.02 \times 10^{-4}c$
with	$2.48 \times 10^{-1}c$	$1.46 \times 10^{-5}c$	$4.24 \times 10^{-6}c$	$2.70 \times 10^{-4}c$	$4.56 \times 10^{-4}c$	$3.05 \times 10^{-4}c$

We simulate the Pb^+ ion beam transport without and with the insulator beam guide. The particle maps of the Pb^+ beam ions and the beam envelope lines $r(z)$ defined by the equation (1) are shown in Fig. 8. For a simulation without the insulator beam guide, the Pb^+ ion distribution in the phase space $(r, P_r/m_0c)$ is shown in Fig. 9(a). Figure 9(b) shows the Pb^+ ion distribution in the phase space $(r, P_r/m_0c)$ of the beam through the insulator guide. The normalized emittance values for the cases without and with the insulator beam guide are 6.41×10^{-6} and 6.06×10^{-6} at $t = 44.7$ ns, respectively. The initial emittance of the Pb^+ ion beam is 4.83×10^{-6} at $t = 10.4$ ns.

It was found that the heavy ion beam considered in this study maintains its beam quality.

SUMMARY

In this work, we studied the intense heavy ion beam transport through an insulator beam guide. Plasma electrons were emitted from the plasma generated on the insulator inner surface. The electrons moved with the heavy ion beam, and the beam charge was neutralized effectively by the electrons. Consequently, the heavy ion beam propagated efficiently through the insulator beam guide. The self-regulated charge neutralization is a unique feature of the simple transport system.

REFERENCES

1. Tabak, M., and Callahan-Miller, D., *Phys. Plasmas* **5**, 1895 (1998).
2. Tabak, M., Callahan-Miller, D., Ho, D. D.-M., and Zimmerman G. B., *Nucl. Fusion* **38**, 509 (1998).
3. Miller, R. B., *An Introduction to the Physics of Intense Charged Particle Beams*, New York: Plenum Press, 1982.

4. Vijayan, T., Roychowdhury, P., and Iyyengar, S. K., *IEEE Trans. Plasma Sci.* **22**, 199 (1994).
5. Miller, H. C., *IEEE Trans. Electr. Insul.* **24**, 765 (1989).
6. Masten, G. B., Ph.D dissertation Texas Tech. Univ. (1993).
7. Kawata, S., Kato, S., Hanamori, S., Nishiyama, S., Naito, K., and Hakoda, M., *Jpn. J. Appl. Phys.* **35**, L1127 (1996).
8. Hanamori, S., Kawata, S., Kato, S., Kikuchi, T., Fujita, A., Chiba, Y., and Hikita, T., *Jpn. J. Appl. Phys.* **37**, L471 (1998).
9. Yamaki, T., "A DESIGN STUDY OF HIF, HIBLIC-I," in *Proceedings of 1984 INS Int. Symp. on Heavy Ion Accelerators and Their Applications to Inertial Fusion*, 1984, pp.141-154.
10. Friedman, A., Bangerter, R. O., and Herrmannsfeldt, W. B., Lawrence Livermore National Laboratory report UCRL-JC-117332, 17 (1994).
11. Birdsall, C. K., and Langdon, A. B., *Plasma Physics via Computer Simulation*, New York: McGraw-Hill Press, 1985.
12. Hockney, R. W., and Eastwood, J. W., *Computer Simulation Using Particles*, New York: McGraw-Hill Press, 1981.
13. Langdon, A. B., and Lasinski, B. F., *Meth. Comp. Phys.* **16**, 327 (1976).

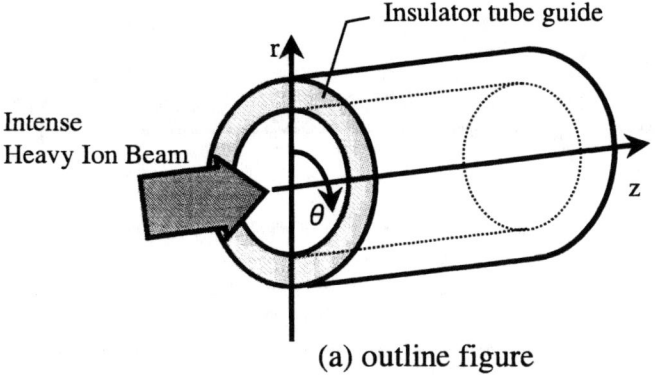

(a) outline figure

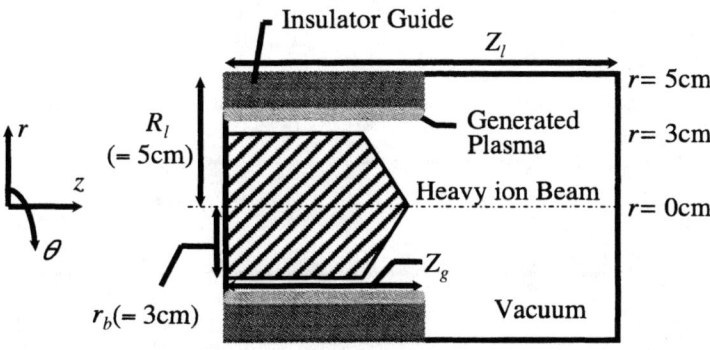

(b) calculation model

FIGURE 1. (a) A system for the intense-heavy-ion-beam transport through the insulator beam guide and (b) simulation model. The transport area length Z_l is 2m or 5m and its radius R_l is 5cm. The beam radius r_b is 3cm. The insulator guide length Z_g is 1m or 2m. The beam body is neutralized by electrons emitted from the guide surface plasma. Local discharges are induced by the electric field, and produce plasma on the insulator inner surface. Electrons are extracted from the plasma by the beam net charge, and neutralize the beam space charge.

(a) Beam ion species: Cs⁺

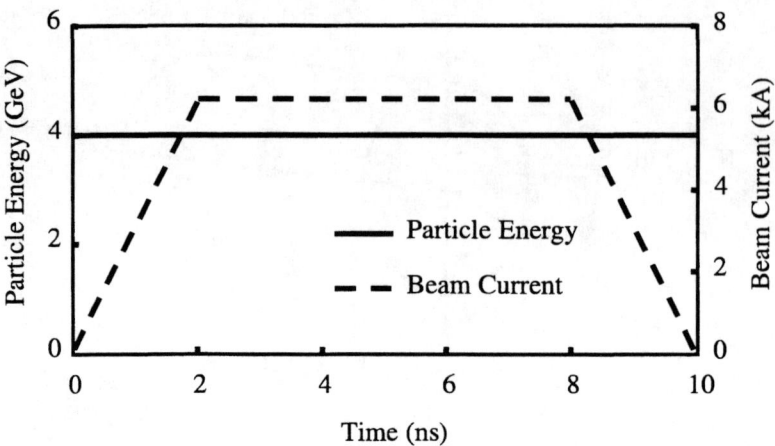

(b) Beam ion species: Pb⁺

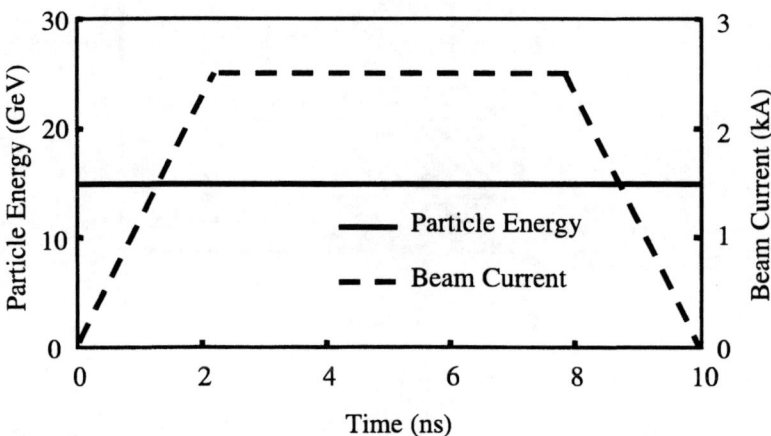

FIGURE 2. (a) Input Cs^+ ion beam waveform: the maximum current I_{0max} is 6.25 kA, the maximum particle energy e_{0max} is 4 GeV, the pulse width is 10 ns, and the rise and fall times are 2 ns. (b) Input Pb^+ ion beam waveform: the maximum current I_{0max} is 2.5 kA, the maximum particle energy e_{0max} is 15 GeV, the pulse width is 10 ns, and the rise and fall times are 2 ns.

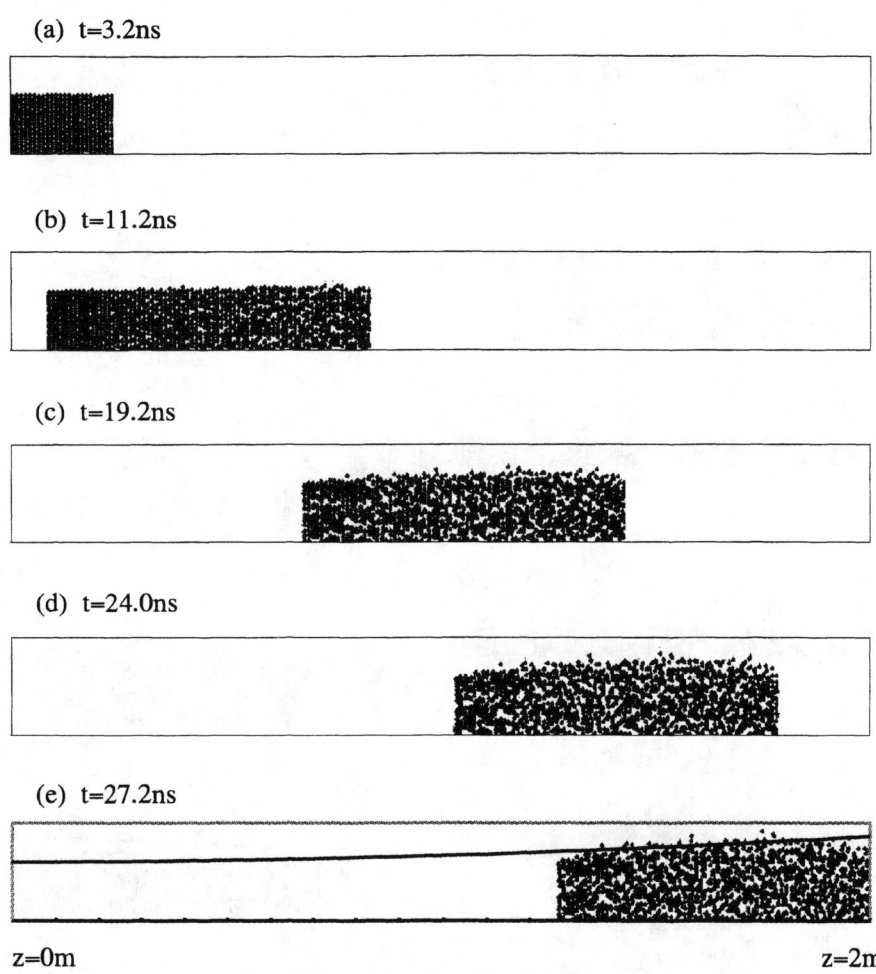

Particle maps for Cs⁺ ion beam

FIGURE 3. Particle maps for the Cs^+ ion beam without the insulator guide: (a) at t=3.2 ns, (b) at t=11.2 ns, (c) at t=19.2 ns, (d) at t=24.0 ns and (e) at t=27.2 ns. The beam radius becomes large by the self space charge. A solid curve in (e) shows the estimate envelope obtained by Eq.(1).

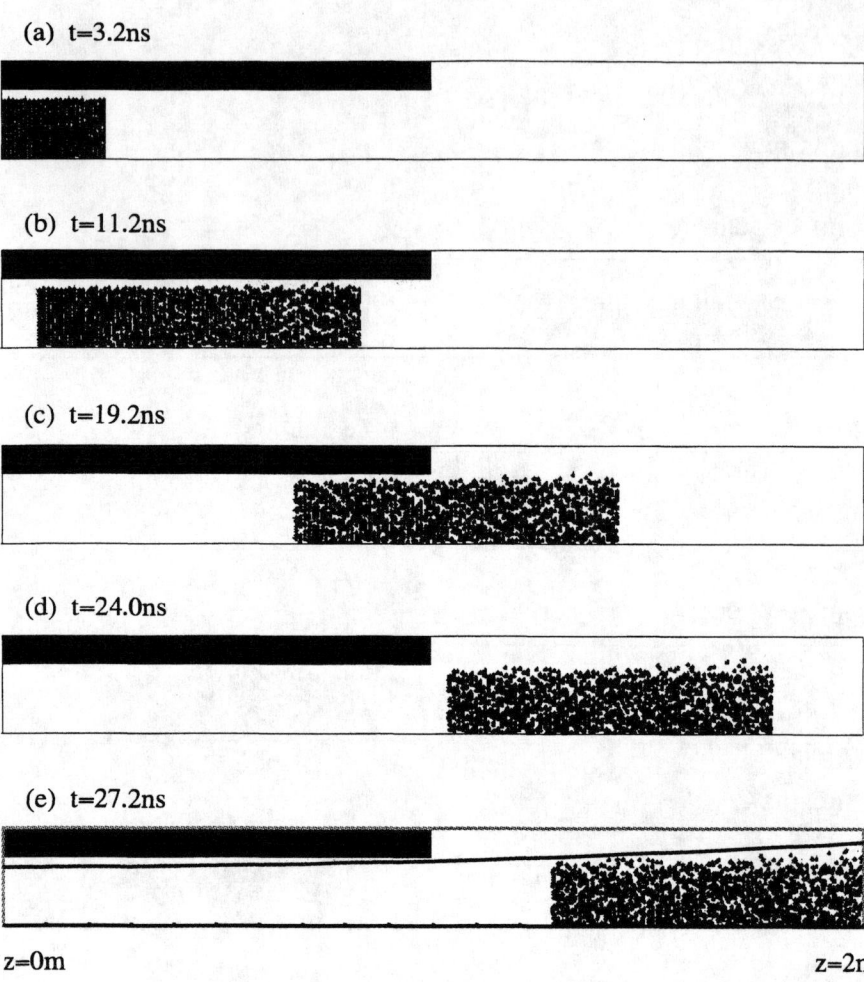

Particle maps for Cs⁺ ion beam

FIGURE 4. Particle maps for the Cs$^+$ ion beam through the insulator beam guide: (a) at t=3.2 ns, (b) at t=11.2 ns, (c) at t=19.2 ns, (d) at t=24.0 ns and (e) at t=27.2 ns. The beam charge is neutralized well by the emitted electrons, and the beam expansion is suppressed. A solid curve in (e) shows the estimate envelope of Cs$^+$ ion beam obtained by Eq.(1), when the beam is not neutralized at all.

Particle maps for emitted electrons

FIGURE 5. Particle maps for the electrons emitted from the insulator beam guide in the case of Cs^+ ion beam transport through the insulator guide: (a) at t=3.2 ns, (b) at t=11.2 ns, (c) at t=19.2 ns, (d) at t=24.0 ns and (e) at t=27.2 ns. The electrons follow the Cs^+ ion beam well.

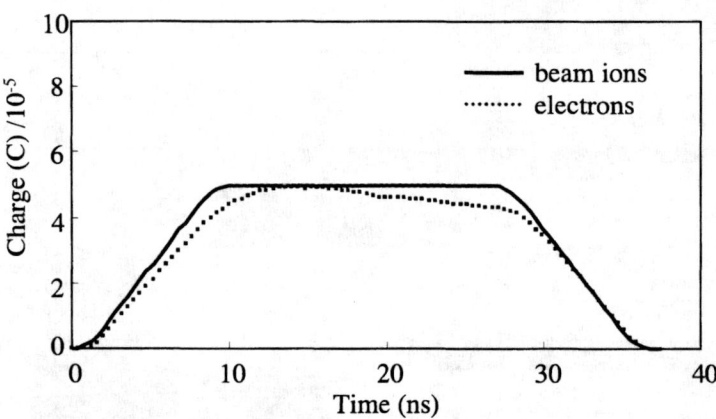

FIGURE 6. History of the total space charges of Cs$^+$ ions and electrons in the transport region. The emitted electrons neutralize the beam net charge. The charge neutralization is self-regulated.

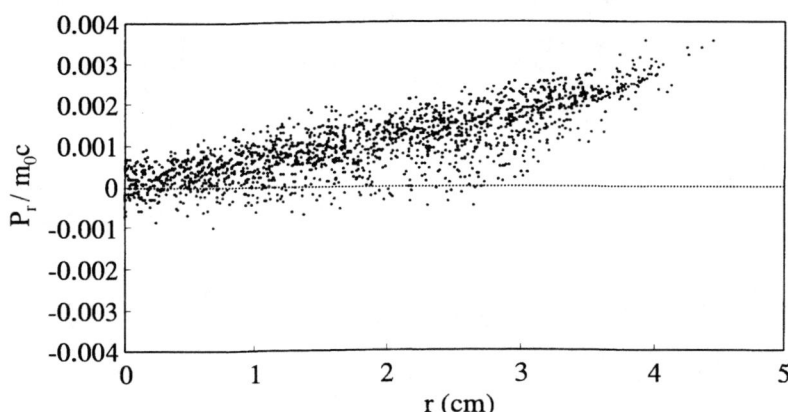

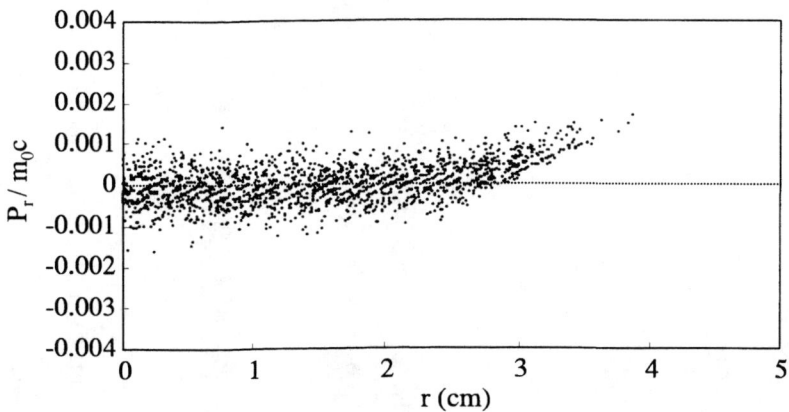

FIGURE 7. The Cs$^+$ ion distribution in the phase space (r, P_r/m_0c) without (a) and with (b) the insulator guide at t=26.8 ns.

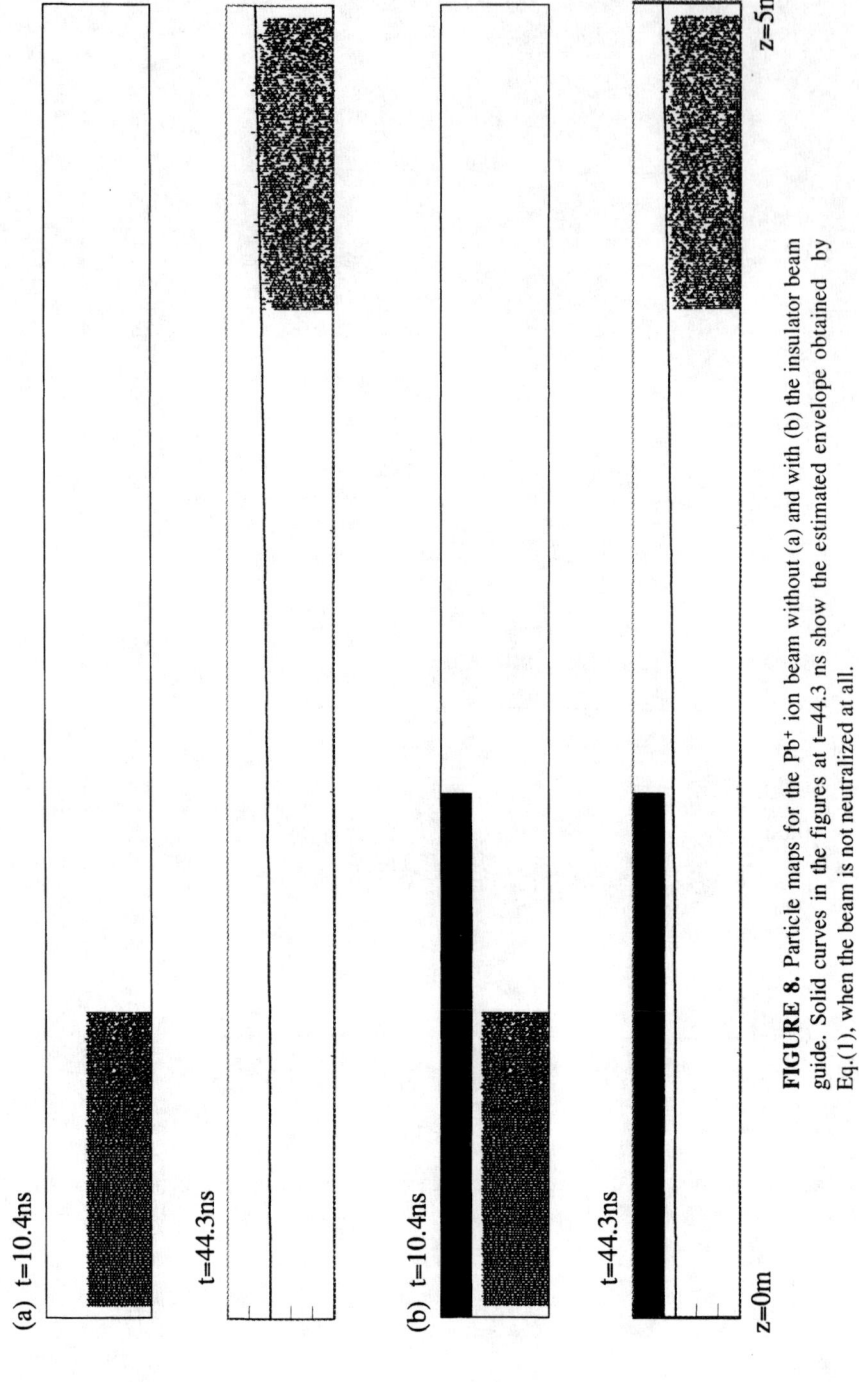

FIGURE 8. Particle maps for the Pb$^+$ ion beam without (a) and with (b) the insulator beam guide. Solid curves in the figures at t=44.3 ns show the estimated envelope obtained by Eq.(1), when the beam is not neutralized at all.

(a) without the insulator guide

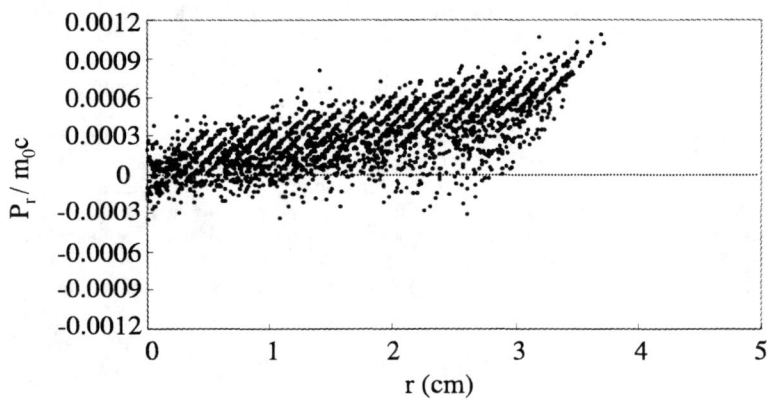

(b) with the insulator guide

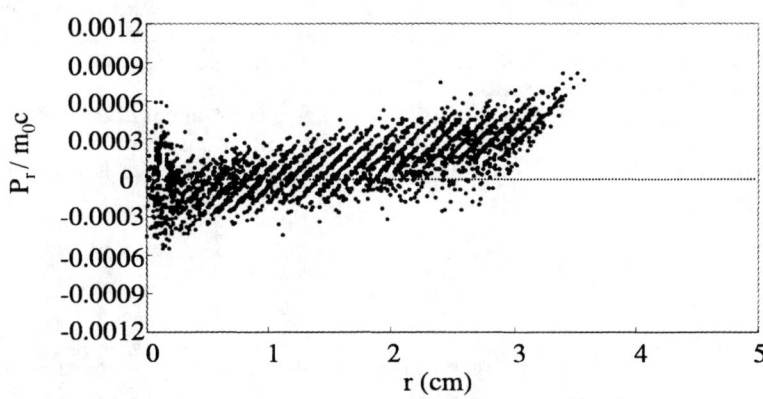

FIGURE 9. The Pb$^+$ ion distribution in the phase space (r, P_r/m_0c) without (a) and with (b) the insulator guide at t=44.7 ns.

Self-Consistent Beam Distribution in Continuous Transport Lines and RF Field

Yuri K. Batygin

The Institute of Physical and Chemical Research (RIKEN), Saitama, 351-01, Japan

Abstract. New self-consistent solutions for beam distribution function in continuos channels with arbitrary applied field are discussed. Both transport and accelerated high brightness beams exhibit the same property: shielding from external field. It gives the universal approach to find stationary self-consistent solution for beam distribution. Particle dynamics in quadrupole channel with higher order components is analyzed. Incorporation of octupole component is not sufficient to suppress beam emittance growth. Better results are obtained utilizing duodecapole component in quadrupole channel. Numerical example of high brightness beam transport with suppressed halo is given. Self-consistent bunch shape in RF field is found which is different from the widely used approximation of bunch by ellipsoid. Equipartitioning in RF field as a result of stationarity of collisionless beam is discussed.

INTRODUCTION

Emittance conservation and prevention of halo formation in a high brightness particle beam are issues for heavy ion fusion accelerator projects. Focusing of intense beam into small spot at the target requires conservation of beam brightness during beam transport and acceleration. If the beam is matched with external focusing and accelerating field, beam distribution function as well as beam emittance are conserved. Matched stationary beam does not exhibit halo formation. Finding matched conditions for the beam requires solutions of the self-consistent problem for beam distribution function in 6-dimensional phase space.

Beam dynamics in a linear focusing channel is a well understood problem. The Kapchinsky - Vladimirsky (KV) distribution is the self- consistent distribution in a linear focusing channel [1]. KV beam in continuos channel obeys KV equation for beam envelope R

$$\frac{d^2R}{dz^2} + \mu_o^2 R - \frac{\varepsilon^2}{(\beta\gamma)^2 R^3} - \frac{2I}{R(\beta\gamma)^3 I_c} = 0, \quad (1)$$

where ε is a normalized beam emittance, β is a particle velocity, γ is a particle energy, I is a beam current, $I_c = 4\pi\varepsilon_o mc^3/q$ is a characteristic value of beam current and μ_o is a transverse particle oscillation frequency in external field with potential U_{ext}:

$$\mu_o^2 = \frac{q}{m\gamma\beta^2 c^2} \frac{1}{r} \frac{dU_{ext}}{dr}. \quad (2)$$

If the beam is in equilibrium with external focusing filed, radius of the beam R is constant. Assuming $d^2R/dz^2 = 0$, envelope equation (1) in this case can be rewritten as an equilibrium condition for KV beam:

$$U_{ext} = \frac{mc^2}{q} \frac{1}{\gamma} (\frac{r}{R})^2 [\frac{1}{2}(\frac{\varepsilon}{R})^2 + \frac{I}{\beta\gamma I_c}] \, , \text{ or for field} \quad (3)$$

$$E_{ext} = -\frac{mc^2}{qR} \frac{1}{\gamma} (\frac{r}{R}) [(\frac{\varepsilon}{R})^2 + 2\frac{I}{\beta\gamma I_c}] \, . \quad (4)$$

Equations (3), (4) contain two terms: one is proportional to square of normalized beam emittance, ε^2, and the second one is proportional to the value of beam current, I. Ratio of that two terms defines criteria for dominance of space charge forces over defocusing due to thermal spread of transverse momentum (beam emittance):

$$b = \frac{2}{\beta\gamma} \frac{I}{\varepsilon^2} \frac{R^2}{I_c} \, . \quad (5)$$

Dimensionless parameter b is a ratio of beam brightness I/ε^2 to normalized value I_c/R^2 and has a meaning of dimensionless beam brightness. Regime b>>1 corresponds to beam transport with dominated space charge forces, while regime b<<1 is a transport with dominance of thermal spread of transverse momentum.

Let us rewrite equation (3) as follow:

$$U_{ext} = \frac{mc^2}{q} \frac{1}{\gamma} (\frac{r}{R})^2 (\frac{I}{\beta\gamma I_c}) (1 + \frac{1}{b}) \, . \quad (6)$$

For high brightness beam transport, the value of 1/b is small with respect to unit. In this case required potential is mainly defined by the value of beam current and practically does not depend on the value of beam emittance.

BEAM EMITTANCE GROWTH AND HALO FORMATION DUE TO BEAM DENSITY REDISTRIBUTION

Realistic beam is not necessary to be uniform. If the beam is essentially non-uniform and propagates in linear focusing channel, it is intrinsically mismatched with the channel. It results in emittance growth and appearance of halo (see Fig. 1). In linear focusing channel non-uniform beam becomes more uniform in the beam core with halo around the beam. Excess of potential energy due to initial mismatching transforms into thermal spread of transverse momentum (free energy process, [2]). Let us give an analytical treatment of beam density redistribution under self non-linear space charge forces in linear focusing channel.

Consider beam with initial Gaussian distribution:

$$f = f_o \exp(-2\frac{x_o^2 + y_o^2}{R^2} - 2\frac{p_{xo}^2 + p_{yo}^2}{(mc)^2 (\frac{\varepsilon}{R})^2}) \, , \quad (7)$$

where $\varepsilon = \frac{4}{mc}\sqrt{<x^2><p_x^2>}$ is a normalized root-mean-square beam emittance. Space charge density is obtained by integration of Eq. (7) over particle momentum:

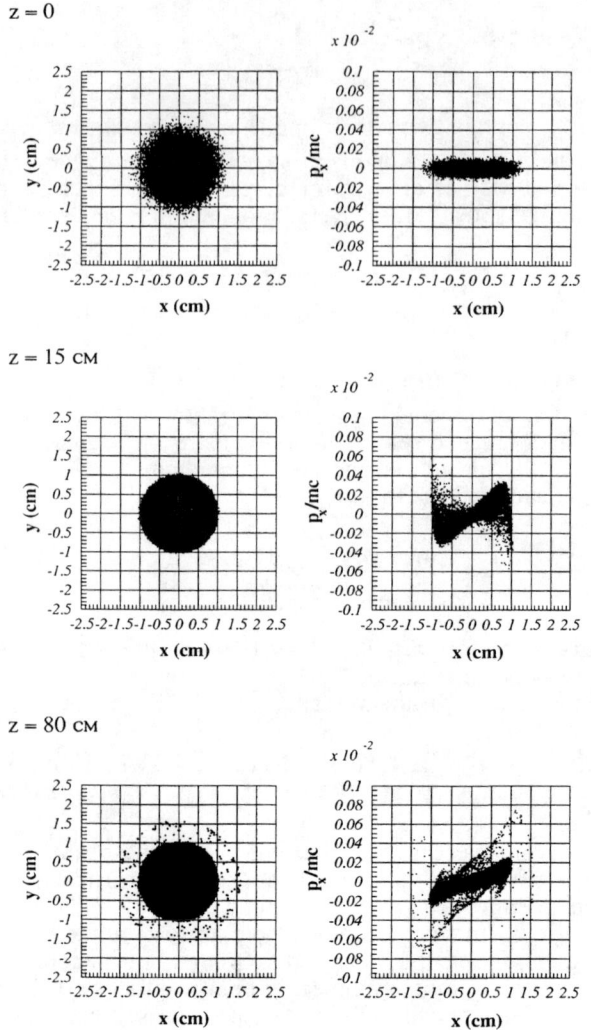

FIGURE 1. Emittance distortion and halo formation of 50 keV, 0.5 A, 0.07 π cm mrad non-uniform proton beam in focusing channel with linear field.

$$\rho(r_o) = \int_{-\infty}^{\infty}\int_{-\infty}^{\infty} f\, dp_{xo}\, dp_{yo} = \rho_o \exp[-2(\tfrac{r_o}{R})^2], \qquad \rho_o = \frac{2I}{\pi c \beta R^2}, \qquad (8)$$

Space charge field of the beam is given by:

$$E_b = \frac{I}{2\pi \varepsilon_o \beta c}\frac{1}{r}[1 - \exp(-2\tfrac{r^2}{R^2})], \qquad (9)$$

At the initial moment of time $t = 0$, total field, E, is a combination of external focusing field E_{ext}, Eq. (4), and space charge field, E_b, Eq. (9):

$$E = E_{ext} + \frac{E_b}{\gamma^2} = \frac{mc^2}{qR\gamma}\frac{2I}{\beta\gamma I_c}(\tfrac{r}{R})[2\frac{1-\exp(-2r^2/R^2)}{(2r^2/R^2)} - 1 - \tfrac{1}{b}], \qquad (10)$$

Let us apply symplectic integrator for particle motion in a combined field, Eq. (10):

$$p_x = p_{xo} + q(\tfrac{x}{r})E t, \qquad p_y = p_{yo} + q(\tfrac{y}{r})E t,$$

$$x = x_o + \frac{p_x}{m\gamma}t, \qquad y = y_o + \frac{p_y}{m\gamma}t. \qquad (11)$$

Inverse transformation is:

$$p_{xo} = p_x - q(\tfrac{x}{r})E t, \qquad p_{yo} = p_y - q(\tfrac{y}{r})E t,$$

$$x_o = x - \frac{p_x}{m\gamma}t, \qquad y_o = y - \frac{p_y}{m\gamma}t. \qquad (12)$$

Substitution of the inverse transformation, Eq. (12) into initial distribution function, Eq. (7), gives the distribution function at the moment t:

$$f = f_o \exp\left(-2\frac{(x - \tfrac{p_x}{m\gamma}t)^2 + (y - \tfrac{p_y}{m\gamma}t)^2}{R^2} - 2\frac{(p_x - q\tfrac{x}{r}E t)^2 + (p_y - q\tfrac{y}{r}E t)^2}{(mc)^2(\tfrac{\varepsilon}{R})^2}\right). \qquad (13)$$

After some algebra Eq. (13) becomes:

$$f = f_o \exp 2\{-(\tfrac{p_x}{mc})^2[(\tfrac{tc}{\gamma R})^2 + (\tfrac{R}{\varepsilon})^2] + 2(\tfrac{p_x}{mc})[(\tfrac{xct}{\gamma R^2}) + \tfrac{x}{r}\tfrac{qEtR^2}{mc\,\varepsilon^2}] - (\tfrac{x}{R})^2 - (\tfrac{x}{r})^2(\tfrac{qEtR}{mc\varepsilon})^2$$

$$- (\tfrac{p_y}{mc})^2[(\tfrac{tc}{\gamma R})^2 + (\tfrac{R}{\varepsilon})^2] + 2(\tfrac{p_y}{mc})[(\tfrac{yct}{\gamma R^2}) + \tfrac{y}{r}\tfrac{qEtR^2}{mc\,\varepsilon^2}] - (\tfrac{y}{R})^2 - (\tfrac{y}{r})^2(\tfrac{qEtR}{mc\varepsilon})^2\}. \qquad (14)$$

Space charge density is obtained by integration of Eq. (14) over particle momentum, utilizing formula [3]:

$$\int_{-\infty}^{\infty} \exp(-p^2 u^2 \pm qu)\, du = \frac{\sqrt{\pi}}{p} \exp\left(\frac{q^2}{4p^2}\right), \qquad p > 0. \qquad (15)$$

After integration, space charge density is

$$\frac{\rho(\xi,\tau)}{\rho_o} = \frac{1}{1+\tau^2} \exp\left\{-\xi^2 - [\tau b F(\xi)]^2 + \frac{\tau^2}{1+\tau^2}[\xi + bF(\xi)]^2\right\} \qquad (16)$$

where the following notations are used:

$$\xi = \sqrt{2}\,\frac{r}{R}, \qquad \tau = t\,\frac{c\,\varepsilon}{\gamma R^2}, \qquad F(\xi) = \xi\left[2\,\frac{1-\exp(-\xi^2)}{\xi^2} - 1 - \frac{1}{b}\right]. \qquad (17)$$

At the moment $t = 0$, space charge density distribution is Gaussian, Eq. (8). Nonlinear space charge forces in linear focusing channel result in uniforming of beam density distribution, Eq. (16). Fig. 2 illustrates beam density redistribution due to space charge forces. Effect of beam uniforming is defined by the product of dimensionless beam brightness b, Eq. (5), and nonlinear function $F(\xi)$, Eq. (17). For more bright beam the more sharp changes in beam density appear.

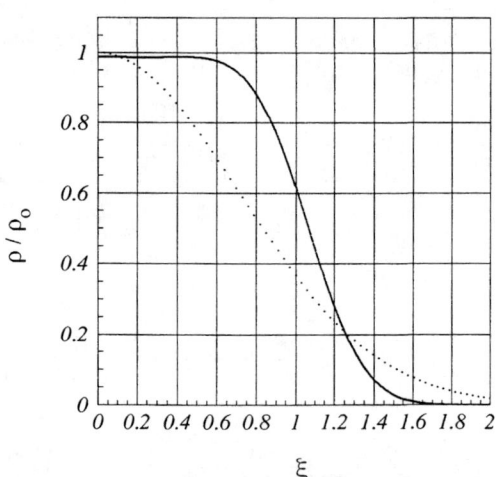

FIGURE 2. Density redistribution of high brightness beam due to non-linear space charge forces, b = 100: dotted line - initial distribution $\tau = 0$, solid line - $\tau = 0.1$, calculated from Eq. (16).

SELF-CONSISTENT BEAM DISTRIBUTION IN CONTINUOUS CHANNEL WITH ARBITRARY FOCUSING FIELD

To match the nonuniform beam with the channel, external focusing field has to be essentially nonlinear [4]. Nonlinear field can be created in a quadrupole channel with higher order components. Before coming to that point, let us discuss the general solution for stationary high-brightness beam distribution in continuous channel with arbitrary applied focusing field [5].

Self-consistent problem for beam distribution in stationary field includes solution of set of two equations: Vlasov's equation for beam distribution function and Poisson's equation for space charge potential of the beam:

$$\begin{cases} \frac{1}{m\gamma}(\frac{\partial f}{\partial x} p_x + \frac{\partial f}{\partial y} p_y) - q(\frac{\partial f}{\partial p_x}\frac{\partial U}{\partial x} + \frac{\partial f}{\partial p_y}\frac{\partial U}{\partial y}) = 0 \\ \\ \frac{1}{r}\frac{\partial}{\partial r}(r\frac{\partial U_b}{\partial r}) + \frac{1}{r^2}\frac{\partial^2 U_b}{\partial \varphi^2} = -\frac{q}{\varepsilon_o}\int_{-\infty}^{\infty}\int_{-\infty}^{\infty} f(x, p_x, y, p_y)\, dp_x\, dp_y \end{cases} \quad (18)$$

General approach to find a self-consistent distribution function for time-independent process is to represent it as a function of Hamiltonian $f = f(H)$:

$$f = f_o \exp(-\frac{H}{H_o}) = f_o \exp(-\frac{p_x^2 + p_y^2}{2m\gamma H_o} - q\frac{U_{ext} + \gamma^{-2}U_b}{H_o}), \quad (19)$$

because this class of solutions automatically obeys Vlasov's equation. Substitution of the distribution function (19) into Poisson's equation provides a nonlinear equation for unknown space charge potential of the beam U_b:

$$\Delta U_b = -\frac{q}{\varepsilon_o}\int_{-\infty}^{\infty}\int_{-\infty}^{\infty} f(\frac{p_x^2 + p_y^2}{2m\gamma} + q U_{ext} + q\frac{U_b}{\gamma^2})\, dp_x\, dp_y. \quad (20)$$

After solving equation (20) for the charge potential of the beam U_b, one can find the self-consistent particle distribution which will be maintained in the focusing channel. Solution of the problem for arbitrary applied focusing field was found in [5]. Let us briefly overview method of solution.

Introduce dimensionless variables:

$$V_{ext} = \frac{q U_{ext}}{H_o}, \quad V_b = \frac{q U_b}{H_o}, \quad \xi = \frac{r}{a}, \quad (21)$$

where a is the radius of the channel. Poisson's equation in cylindrical polar coordinates is

$$\frac{1}{\xi}\frac{\partial V_b}{\partial \xi} + \frac{\partial^2 V_b}{\partial \xi^2} + \frac{1}{\xi^2}\frac{\partial^2 V_b}{\partial \varphi^2} = -\Phi_o \exp-(V_{ext} + V_b\gamma^{-2}), \quad \Phi_o = 16 k \gamma \frac{I}{\beta I_c}(\frac{a}{\varepsilon})^2, \quad (22)$$

where form-factor $k = 1...2$ depends on specific distribution. The unknown potential V_b can be expressed as a Fourier-Bessel series

$$V_b = V_o + \overline{V}_b, \qquad \overline{V}_b = \sum_{n=0}^{\infty} \sum_{m=1}^{\infty} J_n(\upsilon_{nm}\xi)(A_{nm}\cos n\varphi + B_{nm}\sin n\varphi), \qquad (23)$$

where $J_n(x)$ is a Bessel function and υ_{nm} is the m-th root of the equation $J_n(x) = 0$. Expansion (23) satisfies Dirichlet boundary condition at the conductive surface of a round pipe $V_b(a) = V_o$. Constant V_o is defined in such way, that the total potential vanishes at the axis:

$$V_{ext}(0,\varphi) + \frac{\overline{V}_b(0,\varphi)}{\gamma^2} + \frac{V_o}{\gamma^2} = 0. \qquad (24)$$

To find an approximate solution of Poisson's equation, let us take the first term in the near-axis expansion of exponential function:

$$\exp(-V_{ext} - \frac{V_b}{\gamma^2}) \approx 1 - V_{ext} - \frac{V_b}{\gamma^2}. \qquad (25)$$

Substitution of expansion (23) into Poisson's equation with approximation (25) gives:

$$V_o + \sum_{n=0}^{\infty} \sum_{m=1}^{\infty} [1 + \frac{\upsilon_{nm}^2 \gamma^2}{\Phi_o}] J_n(\upsilon_{nm}\xi)(A_{nm}\cos n\varphi + B_{nm}\sin n\varphi) = \gamma^2(1-V_{ext}). \qquad (26)$$

For a high brightness beam $b \gg 1$ the factor in square brackets in Eq. (26) is close to unity:

$$1 + \frac{\upsilon_{nm}^2 \gamma^2}{\Phi_o} = 1 + \frac{\upsilon_{nm}^2}{8bk}(\frac{R}{a})^2 = 1 + \delta, \qquad \delta \approx \frac{1}{bk} \ll 1, \qquad (27)$$

therefore, it can be taken out of the sum (26). With this approximation, the self-consistent space charge dominated beam potential near axis is:

$$V_b = -\frac{\gamma^2}{1+\delta} V_{ext}. \qquad (28)$$

From Eq. (28) it follows that the space charge dominated beam always compensates for the focusing field in the beam core regardless of the applied external focusing potential.

The second approximation to the self-consistent potential V_b can be obtained by taking one more term in the expansion of the exponential function:

$$\exp(-V_{ext} - \gamma^{-2} V_b) \approx 1 - V_{ext} - \gamma^{-2} V_b + \frac{(V_{ext} + \gamma^{-2} V_b)^2}{2}. \qquad (29)$$

Repeating similar derivations resulted in Eq. (28), the second approximation to the space charge potential is defined by

$$V_b = \gamma^2(1+\delta - V_{ext}) - \gamma^2 \sqrt{(1+\delta - V_{ext})^2 - V_{ext}(V_{ext} - 2)}. \qquad (30)$$

The space charge distribution of a matched beam can be derived from Poisson's equation via known space charge potential of the beam

$$\rho_b = -\varepsilon_0 \Delta U_b = \frac{\varepsilon_0}{1+\delta} \gamma^2 \Delta U_{ext} . \qquad (31)$$

From Eq. (31) it follows that the space charge density of high brightness beam is defined by the external focusing potential function U_{ext} and is a weak function of the phase space density of the beam. All information about distribution in phase space is concentrated in small parameter $\delta \ll 1$ which does not affect seriously particle distribution in real space.

Analogous result has been obtained in [6] for linear focusing channel. In the limit of high brightness beam, $b \gg 1$, different particle distributions (KV, Water bag and Gaussian) resulted in uniform distribution in real space. It also follows from Eq. (31) as a specific case for external potential $U_{ext} \sim r^2$. Differentiation of external potential in Eq. (31) gives constant value of space charge distribution $\rho_b = $ const, which corresponds to uniformly populated circle beam.

Performed analysis provides an approach to find a self consistent particle distribution in the channel with arbitrary given focusing potential.

BEAM TRANSPORT IN A QUADRUPOLE FOCUSING CHANNEL WITH DUODECAPOLE FIELD COMPONENT

Consider a uniform four vanes structure with potential

$$U(r,\varphi,t) = (\frac{G_2}{2} r^2 \cos 2\varphi + \frac{G_6}{6} r^6 \cos 6\varphi) \sin \omega_0 t , \qquad (32)$$

where G_2 is a quadrupole gradient, G_6 is a duodecapole component and $\omega_0 = 2\pi c/\lambda$ is an operational frequency. The electrical field of the structure is given by:

$$\vec{E}(r,\varphi,t) = [-\vec{i}_r (G_2 r \cos 2\varphi + G_6 r^5 \cos 6\varphi) + \vec{i}_\varphi (G_2 r \sin 2\varphi + G_6 r^5 \sin 6\varphi)] \sin\omega_0 t . \qquad (33)$$

Particle trajectories in the field (33) can be represented as a combination of a slow variation of particle position with fast oscillations of small amplitude. If phase advance of the particle oscillation per period of field variation is much smaller than 2π, the oscillating field (32) can be replaced by an effective scalar potential of the structure [7]

$$U_{ext}(\vec{r}) = \frac{q}{4 m \gamma} \frac{E_0^2(\vec{r})}{\omega_0^2} , \qquad (34)$$

which describes the averaged motion of particle. For the considered structure the effective potential is:

$$U_{ext}(r,\varphi) = \frac{mc^2}{q} \frac{\mu_0^2}{\lambda^2} [\frac{1}{2} r^2 + a_6 r^6 \cos 4\varphi + \frac{a_6^2}{2} r^{10}], \qquad (35)$$

where μ_o is a smooth transverse oscillation frequency and a_6 is a ratio of field components:

$$\mu_o = \frac{q\,G_2\,\lambda^2}{\sqrt{8}\,\pi\,mc^2\,\sqrt{\gamma}}, \qquad a_6 = \frac{G_6}{G_2}. \tag{36}$$

The effective potential (35) is an axially nonsymmetric and a highly nonlinear function of radius. Equipotential lines $U_{ext}(r,\varphi) = C$ are circles near the axis and are transformed to a 45° skewed square far from the axis (see Fig. 3).

Self-consistent space charge distribution of the beam in the structure with effective potential, Eq. (35), is:

$$\rho = \rho_o\,(1 + 10\,a_6\,r^4\cos 4\varphi + 25\,a_6^2\,r^8), \tag{37}$$

$$\rho_o = \frac{1}{(1 + 5a_6^2 R^8)}\,\frac{I}{\beta c \pi R^2}. \tag{38}$$

Self consistent particle distribution (37) has a 4-folded symmetry (see Fig. 4). Every 45° variation of azimuth angle φ results in a change of the particle distribution from a decreasing to an increasing function of radius and vice versa. To treat the matched beam, it is necessary to bound the beam along equipotential lines. Realistic beam distribution has monotonous decreasing distribution, which can be approximated as:

$$\rho_b = \frac{3\,I}{2\pi c\beta\,R^2}\,(1 - \frac{r^2}{2\,R^2})^2. \tag{39}$$

After truncation along equipotential lines, both distributions, Eqs. (37), (39) are close to each other (see Fig. 4).

In Fig. 5 results of beam matching with distribution, Eq. (39) are presented. Required values of quadrupole gradient and duodecapole components to provide matching of such a beam with the channel are [5]:

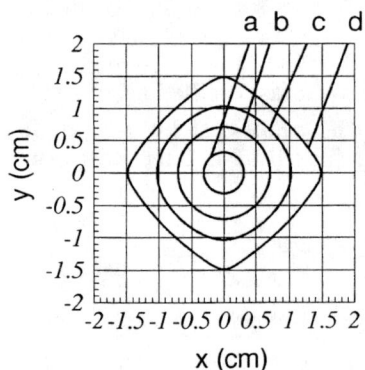

FIGURE 3. Lines of equal values of function $C = \frac{1}{2}r^2 + a_6\,r^6\cos 4\varphi + \frac{a_6^2}{2}r^{10}$ for $a_6 = -0.03$:
(a) $C=0.05$, (b) $C=0.25$, (c) $C=0.5$, and (d) $C=0.85$.

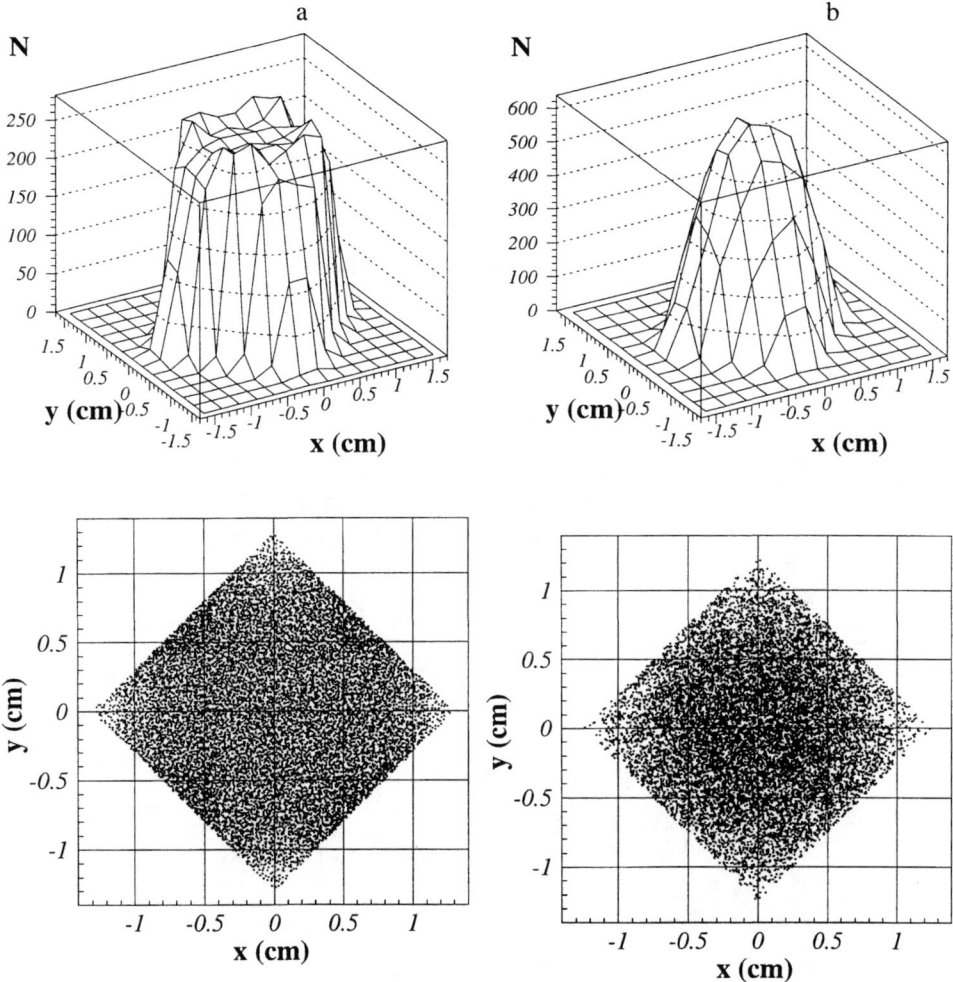

FIGURE 4. (a) Self-consistent particle distribution $\rho_b = \rho_o (1 + 10\zeta r^4 \cos 4\varphi + 25\zeta^2 r^8)$ of the matched beam in a quadrupole channel with a duodecapole component with parameter $\zeta = -0.03$ and (b) beam with distribution $\rho_b = \rho_o [1 - (r/R)^2]^2$, truncated along equipotential lines of external focusing field.

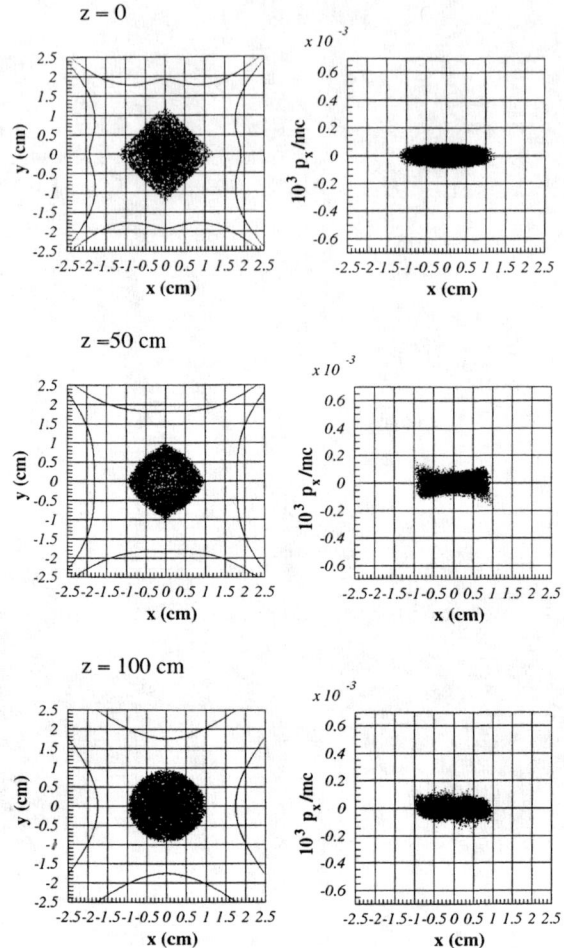

FIGURE 5. Adiabatic matching of the 150 keV, 100 mA, 0.06 π cm mrad proton beam with truncated distribution $\rho_b = \rho_o [1 - (\frac{r}{R})^2]^2$ in quadrupole structure with duodecapole component.

$$G_2 = \frac{\sqrt{8}\,\pi\,m\,c^2}{q\,\lambda\,R}\sqrt{\frac{\varepsilon^2}{R^2} + \frac{3\,I}{I_c\,\beta\gamma}}, \qquad (40)$$

$$G_6 = -\frac{G_2}{12\,\beta\gamma\,R^4}\frac{I}{I_c}\left(\frac{\varepsilon^2}{R^2} + \frac{3\,I}{I_c\,\beta\gamma}\right)^{-1}. \qquad (41)$$

Beam with distribution (39) cannot be exactly matched with the channel, which expresses as a small emittance distortion in phase space. Nevertheless, such a beam is approximately matched with the channel, which gives suppression of halo.

BEAM TRANSPORT IN QUADRUPOLE CHANNEL WITH OCTUPOLE COMPONENT

Let us check how matching conditions are changed if octupole component is incorporated in a quadrupole focusing channel. Consider focusing structure with electrostatic potential

$$U(r,\varphi,t) = \left(\frac{G_2}{2} r^2 \cos 2\varphi + \frac{G_4}{4} r^4 \cos 4\varphi\right) \sin \omega_0 t, \qquad (42)$$

where G_4 is an octupole field component. Oscillating field (42) creates the effective scalar potential

$$U_{ext}(r,\varphi) = \frac{mc^2}{q}\frac{\mu_0^2}{\lambda^2}\left[\frac{1}{2} r^2 + a_4 r^4 \cos 2\varphi + \frac{a_4^2}{2} r^6\right], \qquad (43)$$

where $a_4 = G_4/G_2$ is a ratio of field components. Equipotential lines $U_{ext} = C$ are transformed from circles at small values of radius r, to distorted ellipse at larger r, see Fig. 6.

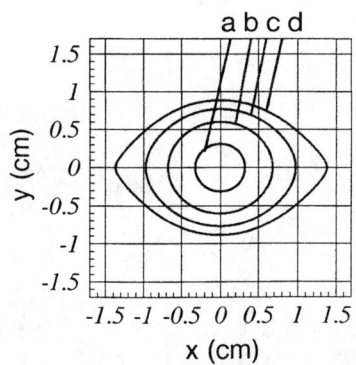

FIGURE 6. Equipotential lines $C = \frac{1}{2} r^2 + a_4 r^4 \cos 2\varphi + \frac{a_4^2}{2} r^6$ for $a_4 = -0.148$
(a) C=0.05, (b) C=0.2, (c) C=0.35, (d) C=0.49.

Space charge density of the matched high-brightness beam in the considered channel is given by:

$$\rho_b(r,\varphi) = \rho_0[1 + 6\, a_4\, r^2\cos2\varphi + 9a_4^2 r^4] \; . \quad (44)$$

Beam with distribution, Eq. (44), is maintained in the channel with potential, Eq. (42). Due to the term $\cos 2\varphi$, space charge density, Eq. (44), is a decreasing function with x-coordinate, but increasing function with y-coordinate, (see Fig. 7a).

Realistic beam distribution is far from matched beam, Eq. (44). Due to monotonic decreasing of beam density distribution, realistic beam is expected to be matched in x-coordinate, but mismatched in y-coordinate.

In Fig. 8 results of simulations of the beam with distribution (39) in a quadrupole channel with octupole component are presented. Initial beam distribution given by Eq. (39) was truncated along equipotential lines to make beam close to the matched beam, Eq. (44). As seen, after 100 cm of beam transport, which corresponds to one betatron particle oscillation, emittance of the beam is slightly distorted at x-p_x phase plane, but seriously distorted at y-p_y phase plane. Therefore, utilizing octupole component in a quadrupole channel is not sufficient to provide beam matching. Better results are observed in computer simulation of the beam in a quadrupole structure with duodecapole field component (see previous Section). In that case the effective potential and matched beam profile are symmetric functions with y as well as x coordinates, and realistic beam is much better matched with focusing structure.

MATCHED BEAM IN QUADRUPOLE CHANNEL WITH HIGHER ORDER COMPONENTS

Including of higher order terms in quadrupole channel makes an analysis much more complicated. Consider structure with potential

$$U = (\frac{G_2}{2} r^2\cos2\varphi + \frac{G_6}{6} r^6\cos6\varphi + \frac{G_{10}}{10} r^{10}\cos10\varphi + \frac{G_{14}}{14} r^{14}\cos14\varphi)\sin\omega_0 t \; . \quad (45)$$

Matched beam distribution is given by

$$\rho(r,\varphi) = \rho_0[1 + 10\, r^4 a_6 \cos4\varphi + r^8(18 a_{10}\cos8\varphi + 25 a_6^2) + 169 r^{24} a_{14}^2$$

$$+ r^{12}(26 a_{14}\cos12\varphi + 90\, a_6\, a_{10}\cos4\varphi) + r^{16}(81\, a_{10}^2 + 130\, a_6\, a_{14}\cos8\varphi)$$

$$+ 234 r^{20} a_{10}\, a_{14}\cos4\varphi]\, , \quad (46)$$

where $a_{10} = G_{10}/G_2$ and $a_{14} = G_{14}/G_2$. Higher order field components result in more round matched beam boundaries, but in the same time in more non-uniform matched beam profile with increasing - decreasing functions of azimuth angle. Realistic beam profile becomes essentially mismatched with the required function, Eq. (46). Incorporating of higher order, than duodecapole, field component in quadrupole channel, does not improve matching of realistic beam with the structure.

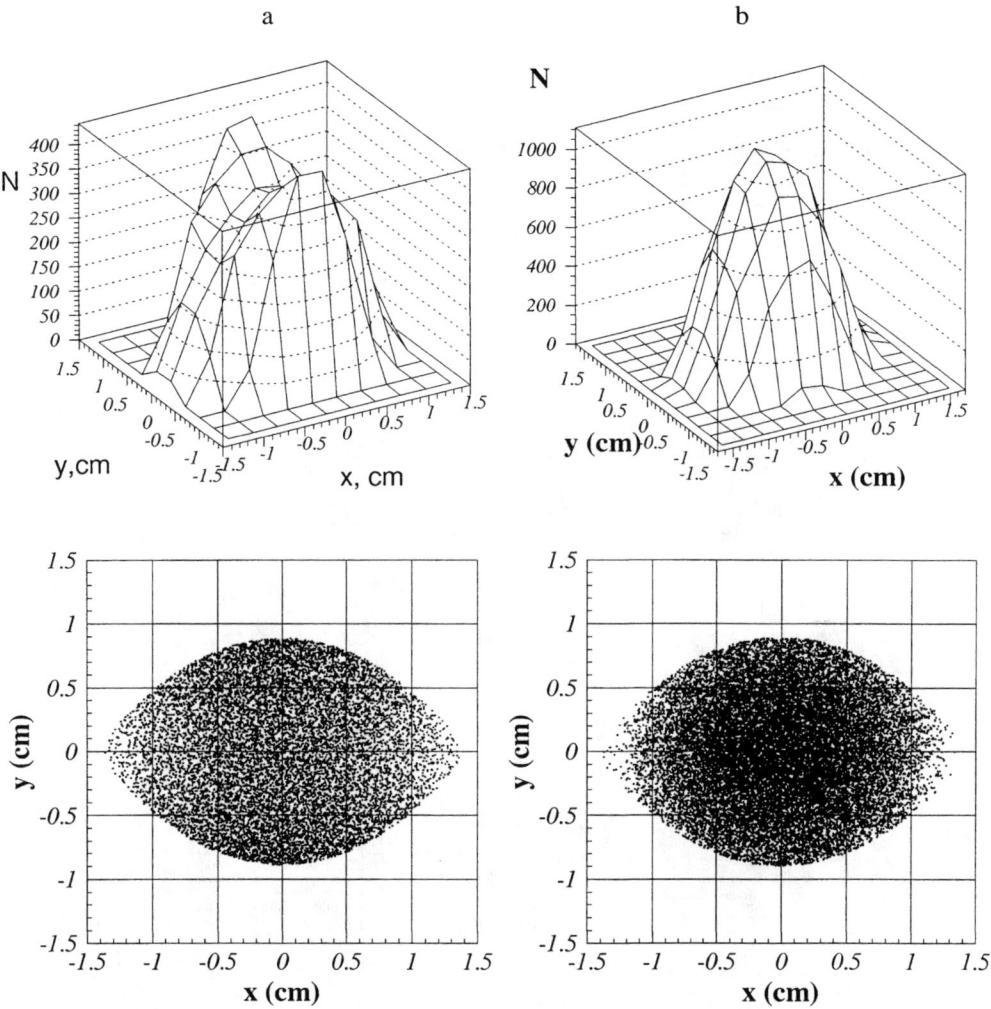

FIGURE 7. (a) Self - consistent beam distribution $\rho_b = \rho_o (1 + 6\eta r^2 \cos 2\varphi + 9\eta^2 r^4)$ in a quadrupole channel with octupole component, $\eta = -0.1$ and (b) beam with distribution $\rho_b = \rho_o [1 - (\frac{r}{R})^2]^2$, truncated along equipotential lines of external focusing field.

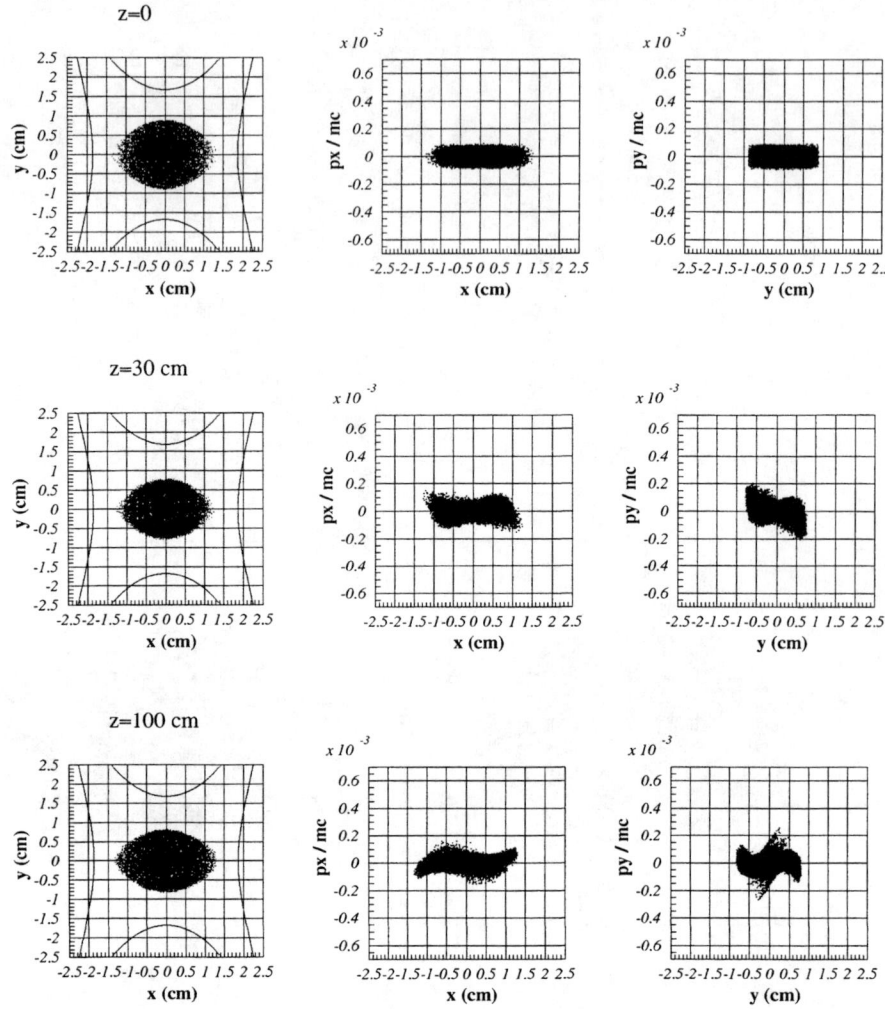

FIGURE 8. Dynamics of the 150 keV, 100 mA, 0.06 π cm mrad proton beam with initial distribution $\rho_b = \rho_o [1 - (\frac{r}{R})^2]^2$ in a four vanes quadrupole structure with field gradient G_2=50kV/cm^2 and octupole component G_4 = -3.5 kV/cm^4.

SELF-CONSISTENT BEAM DISTRIBUTION IN RF FIELD

Consider how matched conditions are changed if RF field is applied. Beam is supposed to be bunched at the frequency $\omega = 2\pi c/\lambda$, where λ is a wavelength. Average longitudinal particle velocity of the beam is $\beta_s = v_s/c$, therefore distance between bunches is $\beta_s \lambda$. Particle motion is governed by single-particle Hamiltonian [6]:

$$H = \frac{p_x^2 + p_y^2}{2m\gamma} + \frac{p_z^2}{2m\gamma^3} + q U_{ext} + q \frac{U_b}{\gamma^2}, \tag{47}$$

$$U_{ext} = \frac{q v_s E}{\omega} [I_o(\frac{\omega r}{\gamma v_s}) \sin(\varphi_s - \frac{\omega}{v_s}\zeta) - \sin\varphi_s + \frac{\omega}{v_s}\zeta \cos\varphi_s] + m\gamma\Omega_r^2 \frac{r^2}{2}, \tag{48}$$

where $p_z = p - p_s$ is deviation from longitudinal momentum, $\zeta = z - v_s t$ is a longitudinal deviation from synchronous particle, E is an amplitude of accelerating field, φ_s is a synchronous phase, Ω_r is a transverse oscillation frequency without space charge forces and accelerating field.

Particle distribution of a moving bunched beam has a form of $\rho = \rho(x, y, z - v_s t)$. Moving charge distribution creates an electromagnetic field with scalar potential $U_b = U_b(x, y, z - v_s t)$ and vector potential $\vec{A}_b = \vec{A}_b(x, y, z - v_s t)$, which obey wave equations [8]:

$$\Delta U_b - \frac{1}{c^2}\frac{\partial^2 U_b}{\partial t^2} = -\frac{\rho}{\varepsilon_o},$$

$$\Delta \vec{A}_b - \frac{1}{c^2}\frac{\partial^2 \vec{A}_b}{\partial t^2} = -\mu_o \vec{j}, \tag{49}$$

where $\vec{j} = \rho \vec{v}_s$ is a current density of the beam. Vector $\vec{j}$ has only longitudinal component

$$j_x = j_y = 0, \qquad j_z = v_s \rho(x, y, z - v_s t), \tag{50}$$

and, therefore, vector potential has only longitudinal component A_z. In a moving coordinate system, where particles are static, the vector potential of the beam is zero $\vec{A} = 0$. According to Lorentz transformation, component of vector potential in laboratory system is $A_z = \beta_s U_b / c$. Therefore, for solution of the problem of self field of the bunch it is enough to solve only equation for scalar potential:

$$\frac{\partial^2 U_b}{\partial x^2} + \frac{\partial^2 U_b}{\partial y^2} + \frac{\partial^2 U_b}{\gamma^2 \partial \zeta^2} = -\frac{1}{\varepsilon_o}\rho(x, y, \zeta). \tag{51}$$

Equation (51) has to be solved together with Vlasov's equation for beam distribution function:

$$\frac{df}{dt} = \frac{1}{m\gamma}\left(\frac{\partial f}{\partial x}p_x + \frac{\partial f}{\partial y}p_y + \frac{\partial f}{\partial \zeta}p_z\right) - q\left(\frac{\partial f}{\partial p_x}\frac{\partial U}{\partial x} + \frac{\partial f}{\partial p_y}\frac{\partial U}{\partial y} + \frac{\partial f}{\partial p_z}\frac{\partial U}{\partial \zeta}\right) = 0, \quad (52)$$

where $U = U_{ext} + \gamma^{-2}U_b$ is the total potential of the structure.

Beam Equipartitioning in RF Field

Method of solution of self-consistent system of equations (51), (52) is similar to that for transport channel. Beam distribution function is expressed as a function of Hamiltonian $f = f_o \exp(-H/H_o)$:

$$f = f_o \exp\left(-\frac{p_x^2 + p_y^2}{2m\gamma H_o} - \frac{p_z^2}{2m\gamma^3 H_o} - q\frac{U_{ext} + U_b\gamma^{-2}}{H_o}\right). \quad (53)$$

Rewrite distribution function (53) as follow

$$f = f_o \exp\left(-2\frac{p_x^2 + p_y^2}{p_t^2} - 2\frac{p_z^2}{p_l^2} - q\frac{U_{ext} + U_b\gamma^{-2}}{H_o}\right), \quad (54)$$

where $p_t = 2\sqrt{<p_x^2>} = 2\sqrt{<p_y^2>}$ and $p_l = 2\sqrt{<p_z^2>}$ are double root-mean-square beam sizes in phase space. Transverse, ε_t, and longitudinal, ε_l, rms beam emittances are:

$$\varepsilon_t = 2\frac{p_t}{mc}\sqrt{<x^2>} = 2\frac{p_t}{mc}\sqrt{<y^2>}, \quad (55)$$

$$\varepsilon_l = 2\frac{p_l}{mc}\sqrt{<\zeta^2>}. \quad (56)$$

Taking together Eqs. (53) - (56), the value of H_o can be expressed as a function of beam parameters:

$$16 \cdot H_o = \frac{mc^2}{\gamma}\frac{\varepsilon_t^2}{<x^2>} = \frac{mc^2}{\gamma}\frac{\varepsilon_t^2}{<y^2>} = \frac{mc^2}{\gamma^3}\frac{\varepsilon_l^2}{<\zeta^2>}. \quad (57)$$

Equation (57) can be rewritten as

$$\frac{\varepsilon_t}{R} = \frac{\varepsilon_l}{\gamma l}, \quad (58)$$

where $R = 2\sqrt{<x^2>}$ is a beam radius and $l = 2\sqrt{<\zeta^2>}$ is a half-size of the bunch length. Equations (57) and (58) express the equipartitioning condition for the beam in RF field [9]. From the above derivations it is clear, that equipartitioning is a consequence of a stationarity of the beam distribution function. Equation (53) is a solution of Vlasov's equation (52), therefore equipartitioning condition (58) is valid for collisionless beam and is not connected with thermalization process.

Equipartitioning is a necessary condition, i.e. if distribution function is stationary (time independent), equipartitioning is fulfilled. Opposite statement is not valid in general case: there are infinitely large number of distribution functions which obeys condition (58), but are not stationary. Therefore, equipartitioning is not sufficient condition. To find the stationary distribution function it is necessary to solve nonlinear wave equation for unknown potential of the beam.

Space Charge Field of the Bunch

Space charge density of the beam is:

$$\rho(x,y,\zeta) = q \int_{-\infty}^{\infty} \int_{-\infty}^{\infty} \int_{-\infty}^{\infty} f \, dp_x \, dp_y \, dp_z = \rho_o \exp\left(-q \frac{U_{ext} + U_b \gamma^{-2}}{H_o}\right), \quad (59)$$

where ρ_o is the space charge density in the center of the bunch. The value of ρ_o is unknown at this point due to the unknown space charge potential of the beam, U_b. Let us introduce an average value of space charge density, which is equal to the density of an equivalent cylindrical bunch with the same beam radius and the same half-bunch length:

$$\overline{\rho} = \frac{I \lambda}{2\pi R^2 \, l \, c}. \quad (60)$$

The value of ρ_o differs from the average value of space charge density $\overline{\rho}$ as a form-factor k:

$$\rho_o = k \overline{\rho}. \quad (61)$$

Introducing dimensionless variables:

$$V_{ext} = \frac{q \, U_{ext}}{H_o}, \quad V_b = \frac{q \, U_b}{H_o}, \quad \xi = \frac{r}{a}, \quad \eta = \frac{\zeta}{a}, \quad (62)$$

where a is a channel radius, the wave equation (51) in cylindrical polar coordinates becomes

$$\frac{1}{\xi} \frac{\partial V_b}{\partial \xi} + \frac{\partial^2 V_b}{\partial \xi^2} + \frac{\partial^2 V_b}{\partial \eta^2 \gamma^2} = -\frac{8 \, k \, b}{\delta \varphi} \left(\gamma \frac{a}{R}\right)^2 \exp - \left(V_{ext} + \frac{V_b}{\gamma^2}\right). \quad (63)$$

Here, $\delta\varphi \equiv \Delta\varphi/(2\pi)$ is a reduced bunch width and $\Delta\varphi$ is a phase bunch length. To solve the nonlinear equation (63), the same method as for transport channel is applied.

Represent unknown space charge potential of the beam by Fourier-Bessel series:

$$V_b = V_o + \sum_{n=0}^{\infty} \sum_{m=1}^{\infty} J_o(\upsilon_{om} \xi)[A_{nm} \cos(k_z n \eta) + B_{nm} \sin(k_z n \eta)], \quad (64)$$

where $k_z \equiv \omega a / v_s$ is a wave number. Expansion (64) obeys Dirichlet boundary condition $V_b(a) = V_o$ at the perfect conductive surface of the channel and takes into account periodic function of potential due to train of the bunches. Constant V_o is defined in such a way that the total potential of the structure vanishes at the bunch center: $V_{ext}(0,0) + V_b(0,0)\gamma^{-2} = 0$.

To find an approximate solution of equation for scalar potential, let us take only the first term in the near-center expansion of exponential function as $\exp(-V_{ext} - V_b\gamma^{-2}) \approx 1 - V_{ext} - V_b\gamma^{-2}$. Equation (63) then becomes:

$$\sum_{n=0}^{\infty}\sum_{m=1}^{\infty} [1 + \frac{v_{om}^2 + (k_z n)^2 \gamma^{-2}}{8kb}(\delta\varphi)(\frac{R}{a})^2] J_o(v_{om}\xi) [A_{nm}\cos(k_z n\eta) + B_{nm}\sin(k_z n\eta)]$$

$$= (1 - V_{ext})\gamma^2 - V_o . \quad (65)$$

Space charge dominated beam transport is achieved, if $b \gg 1$. It gives a possibility to simplify the equation (65). Expression in square brackets in Eq. (65) is

$$1 + \frac{v_{om}^2 + (k_z n)^2 \gamma^{-2}}{8kb}(\delta\varphi)(\frac{R}{a})^2 = 1 + \delta , \quad (66)$$

$$\delta = \frac{v_{om}^2 + (k_z n)^2 \gamma^{-2}}{8kb}(\delta\varphi)(\frac{R}{a})^2 . \quad (67)$$

Roots of the Bessel function are $v_{o1} = 2.408$, $v_{o2} = 5.52$. Parameter k_z is close to unity:

$$k_z = 2\pi(\frac{a}{\lambda\beta}) \approx 1 . \quad (68)$$

Taking into account, that $\delta\varphi \leq 1$, $R/a \approx 0.5$, the value of δ is much smaller than unity for a high brightness beam:

$$\delta \approx \frac{1}{bk} \ll 1 . \quad (69)$$

Therefore, expression (66), can be taken out of the sum in Eq. (65). With this approximation, self-consistent space charge dominated beam potential is:

$$V_b = -\frac{\gamma^2}{1+\delta} V_{ext} . \quad (70)$$

Second approximation to the self-consistent potential is given by holding one more term in expansion of exponential function and results in the same expression as equation (30). High brightness bunched beam in RF field exhibits the same property as unbunched beam in transport channel, namely shielding itself from the external field.

Taking the first approximation to the space charge potential of the beam (70), the Hamiltonian corresponding to the self-consistent bunch distribution is:

$$H = \frac{p_x^2 + p_y^2}{2m\gamma} + \frac{p_z^2}{2m\gamma^3} + q\left(\frac{\delta}{1+\delta}\right) U_{ext}. \tag{71}$$

Equation (71) indicates that in the presence of intense, bright bunched beam ($\delta \ll 1$) the stationary longitudinal phase space of the beam becomes narrow in momentum spread, remaining, in the first approximation, the same in coordinate. This is in a qualitative agreement with the study of Ref. [6].

Stationary Bunch Profile

Self consistent space charge distribution of matched beam in the channel can be obtained from equation for scalar potential:

$$\rho(r,\zeta) = -\varepsilon_0 \left[\frac{1}{r}\frac{\partial}{\partial r}\left(r\frac{\partial U_b}{\partial r}\right) + \frac{\partial^2 U_b}{\gamma^2 \partial \zeta^2}\right]. \tag{72}$$

Substitution of Eq. (70) into Eq. (72) gives the stationary particle density distribution inside the bunch:

$$\rho(r,\zeta) = \rho_0 \left\{ 1 - \frac{\delta}{\sqrt{(1+\delta)^2 - 2\delta V_{ext}}} - \frac{\delta^2}{32\gamma^2} \frac{\varepsilon_t^2}{\langle x^2 \rangle} \left(\frac{c}{a\Omega_r}\right)^2 \frac{\left(\frac{\partial V_{ext}}{\partial \xi}\right)^2 + \left(\frac{\partial V_{ext}}{\gamma \partial \eta}\right)^2}{[(1+\delta)^2 - 2\delta V_{ext}]^{3/2}} \right\}. \tag{73}$$

For high brightness beam, parameter $\delta \ll 1$, therefore, space charge density is close to constant within the bunch.

From Eq. (70) it follows, that, in the first approximation, space charge potential of the beam is the same function of coordinates, as the external potential, with opposite sign. Therefore, equation $U_{ext}(r, \zeta) = $ const gives the family of equipotential lines of space charge field of the beam:

$$I_0\left(\frac{\omega r}{\gamma v_s}\right)\sin\left(\varphi_s - \frac{\omega \zeta}{v_s}\right) - \sin\varphi_s + \frac{\omega \zeta}{v_s}\cos\varphi_s + C\, r^2 = \text{const}, \tag{74}$$

$$C = \frac{m\gamma \Omega_r^2 \omega}{2 v_s E}. \tag{75}$$

Consider bunch with boundary $R(\zeta)$, defined by nonlinear equation:

$$I_0\left(\frac{\omega R}{\gamma v_s}\right)\sin\left(\varphi_s - \frac{\omega \zeta}{v_s}\right) - \sin\varphi_s + \frac{\omega \zeta}{v_s}\cos\varphi_s + C_1 R^2 = \text{const}. \tag{76}$$

In general case, bunch boundary $R(\zeta)$ does not create an equipotential surface. Nevertheless, space charge potential of uniformly populated bunch with boundary, Eq. (76), is close to that, given by Eq. (48) (see below). The value of constant in Eq. (76) can be defined from the condition, that longitudinal bunch size is, in the first approximation, the same as for zero - current mode. Therefore, at $R(\zeta)=0$, the left bunch boundary is $\zeta = -2\varphi_s$ and the value of constant is

$$\text{const} = 2\varphi_s \cos\varphi_s - 2\sin\varphi_s \ . \tag{77}$$

Substitution of Eq. (77) into Eq. (76) gives the first approximation to the beam profile:

$$F(R,\zeta) = I_o(\frac{\omega R}{\gamma v_s}) \sin(\varphi_s - \frac{\omega\zeta}{v_s}) + \sin\varphi_s - (2\varphi_s - \frac{\omega\zeta}{v_s})\cos\varphi_s + C_1 R^2 = 0 \ . \tag{78}$$

Let us find a potential at the axis, created by uniformly populated bunch with boundary, Eq. (78). Consider for simplicity a non-relativistic case. Potential of arbitrary charge distribution at the point ζ_o at the axis is:

$$U_b(\zeta_o, 0) = \frac{1}{4\pi\varepsilon_o} \int \frac{\rho \, dV}{|\vec{r}|} = \frac{1}{4\pi\varepsilon_o} \int_0^{R(\zeta)} \int_0^{2\pi} \int_{\zeta_{min}}^{\zeta_{max}} \frac{\rho(r,\zeta') \, r \, dr \, d\zeta' \, d\varphi}{\sqrt{r^2 + (\zeta' - \zeta_o)^2}} \ . \tag{79}$$

Let us change from longitudinal coordinate ζ to RF phase $\psi = -\zeta\omega/v_s$. After integration in Eq. (79) over radius and azimuthal angle, the potential is:

$$U_b(\psi_o, 0) = U_o \int_{\psi_{min}}^{\psi_{max}} [\sqrt{k_z^2 R^2(\psi) + (\psi - \psi_o)^2} - \sqrt{(\psi - \psi_o)^2}] \, d\psi \ . \tag{80}$$

For a long bunch, $l \gg R$, it can be assumed $I_o(\omega R/\gamma v_s) \approx 1$. Then the bunch boundary $R(\psi)$ is defined by equation

$$(k_z R)^2 \approx \frac{1}{C_1} [(\psi + 2\varphi_s)\cos\varphi_s - \sin\varphi_s - \sin(\psi + \varphi_s)] \ . \tag{81}$$

Substitution of Eq. (81) into Eq. (80) gives the potential at the axis point ψ_o:

$$U_b(\psi_o, 0) = U_o \int_{\psi_{min}}^{\psi_{max}} [\sqrt{\psi^2 + \psi(\frac{\cos\varphi_s}{C_1} - 2\psi_o) - \frac{\sin(\varphi_s+\psi)}{C_1} + (\frac{2\varphi_s\cos\varphi_s}{C_1} - \frac{\sin\varphi_s}{C_1} + \psi_o^2)}$$

$$- \sqrt{(\psi - \psi_o)^2}] \, d\psi \ . \tag{82}$$

In Fig. 9 the space charge potential of the bunch U_b, Eq. (82), as a function of phase, ψ, is presented. Also, inverse potential of external field at the axis, Eq. (48), is presented:

$$V(\psi) = -U_{ext}(\psi, 0) = -\frac{E}{k_z}[\sin(\psi + \varphi_s) - \psi\cos\varphi_s] \ . \tag{83}$$

As seen, the values are close to each other, which indicates the good approximation of the bunch boundary by Eq. (76).

In Fig. 10 a uniformly populated bunch with boundary, Eq. (78) is presented. As seen, bunch boundary in real space is close to separatrix shape. Space charge forces of the bunch in longitudinal direction are essentially nonlinear and repeat (with negative sign) the RF field inside the bunch. Therefore, uniformly populated bunch with boundary, Eq. (78), compensates for "restoring" self-phasing force inside the bunch. In transverse direction the space charge forces are close to linear and compensate for external focusing forces.

In Fig. 11 numerical results of accelerated particle distribution in high brightness linac with solenoid focusing are presented. Bunch shape is close to that derived by analytical methods of this section.

Finally, let us discuss applicability of the well known approximation of bunch by uniformly populated ellipsoid. In derivations of self-consistent solution of beam distribution presented in this paper there is no any constraints on external potential. Equation (70) is valid for arbitrary external field. Let us consider particle motion in the vicinity of synchronous particle, utilizing expansion

$$\sin(\varphi_s - \frac{\omega}{v_s}\zeta) \approx \sin\varphi_s - (\frac{\omega}{v_s}\zeta)\cdot\cos\varphi_s - \frac{1}{2}(\frac{\omega}{v_s}\zeta)^2 \sin\varphi_s, \quad \frac{\omega\zeta}{v_s} \ll 1. \quad (84)$$

Approximation (84) is valid for longitudinal particle oscillations, much smaller than separatrix size. Additionally, consider radial deviation much smaller than bunch period $r \ll \beta_s\lambda$ so we can assume $I_o(\frac{\omega r}{\gamma v_s}) \approx 1$. Under that restrictions, potential (48) becomes:

$$U_{ext}(r \ll \beta_s\lambda, \frac{\omega\zeta}{v_s} \ll 1) = \frac{1}{2}\frac{q\omega E|\sin\varphi_s|}{v_s}\zeta^2 + \frac{1}{2}m\gamma\Omega_r^2 r^2. \quad (85)$$

Substitution of Eq. (85) into Eq. (70) gives bunch potential, which can be written as a potential of uniformly populated ellipsoid:

$$U_b = -\frac{\rho_o}{\varepsilon_o}(M_z\zeta^2 + \frac{1-M_z}{2}r^2). \quad (86)$$

where $M_z = M_z(R/l)$ is a function of ratio of ellipsoid semi-axes. Therefore, approximation of bunch shape by ellipsoid is valid for small sizes bunch $R \ll \beta_s\lambda$, $l \ll \beta_s\lambda$, which occupies central part of separatrix around synchronous particle.

SUMMARY

In this paper self-consistent solutions for high brightness beam in continuous channels with arbitrary external potential are derived. Analysis is done for unbunched and bunched beams. Both cases exhibit the same property of high brightness beams, namely shielding from external field regardless the external applied potential. It provides an approach to find the self-consistent particle distribution. Performed analysis describes stationary conditions for the beam transport and acceleration without emittance growth and halo formation.

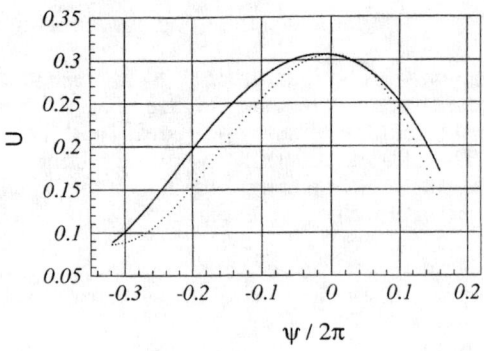

FIGURE 9. Space charge potential $U_b(\psi, 0)$ of the approximate self-consistent bunched beam distribution at the axis, $\varphi_s = -1$, $C_1 = 3.8$; (Solid line), inverse external potential at the axis - $U_{ext}(\psi, 0)$ (dotted line).

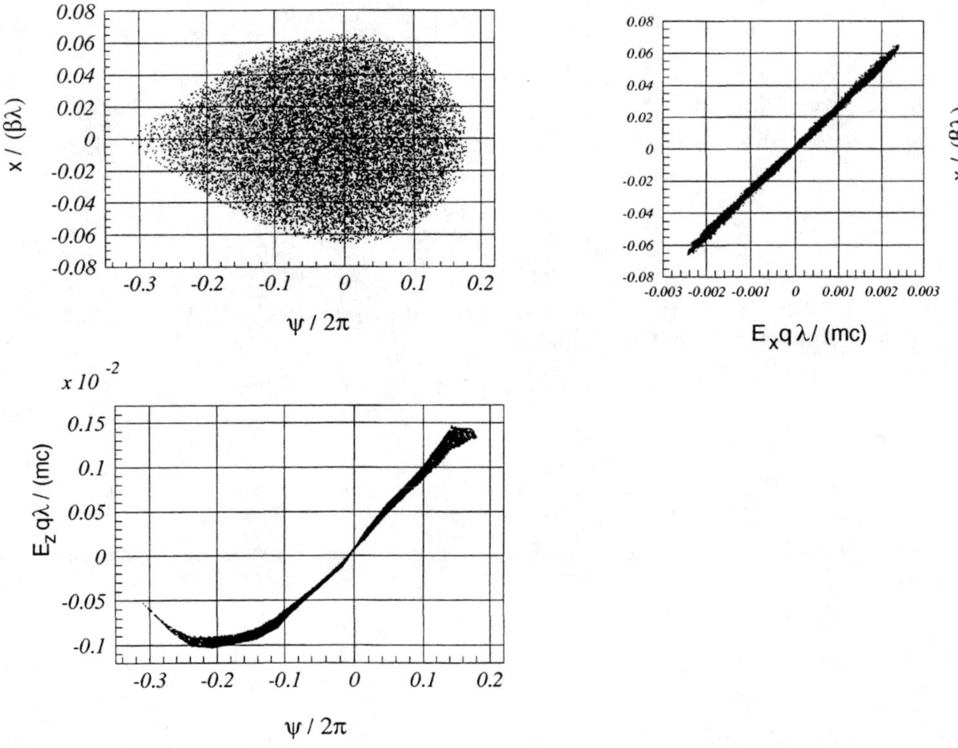

FIGURE 10. Approximate stationary self-consistent particle distribution in RF field and space charge forces of the bunch, $\varphi_s = -1$, $C_1 = 3.8$.

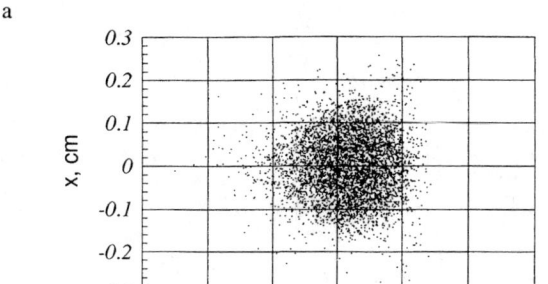
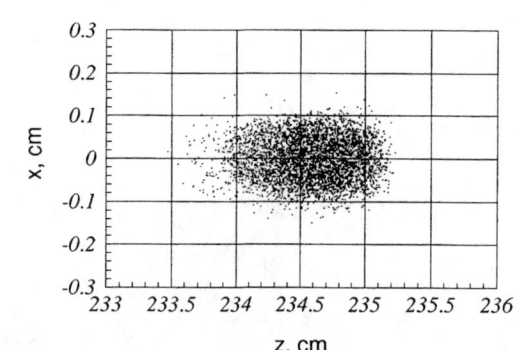

FIGURE 11. Profile of 3 MeV, 250 mA proton bunch in a drift tube linac with solenoid focusing: a) $\frac{\Omega_r}{\omega} = 1.25$, b) $\frac{\Omega_r}{\omega} = 1.75$, c) $\frac{\Omega_r}{\omega} = 2$.

REFERENCES

1. I.M.Kapchinsky, V.V.Vladimirsky, *Proc. of the International Conference on High Energy Acceleration and Instrumentation*, Geneve, 1959, p. 274.
2. M.Reiser, *Theory and Design of Charged Particle Beams,* Wiley, New York, 1994.
3. I.S.Gradshtein and I.M.Ryzhik, *Table of Integrals, Series, and Products,* Academics Press, 1965.
4. Y.Batygin, *Physical Review E*, Vol. 53, p.p. 5358-5365, 1996.
5. Y.Batygin, *Physical Review E*, Vol. 57, p.p. 6020-6029, 1998.
6. I.M.Kapchinsky, *Theory of Resonance Linear Accelerators*, Harwood, 1985.
7. L.D.Landau and E.M.Lifshitz, *Mechanics,* Pergamon Press, 1960.
8. L.D.Landau and E.M.Lifshitz, *The Classical Theory of Field*, Pergamon Press, 1960.
9. R.Jameson, *IEEE Trans. Nucl. Sci.*, NS-28, 2408 (1981).

Particle dynamics in a DTL for high intensity heavy ion beams for inertial fusion[1]

Giovanni Parisi*, Horst Deitinghoff*, Klaus Bongardt[†] and Michael Pabst[†]

* *Institut für Angewandte Physik der J.W.Goethe-Universität*
Robert-Mayer-Strasse 2-4, D-60054 Frankfurt am Main, Germany
[†]*Forschungszentrum Jülich GmbH, Postfach 1913, D-52425 Jülich, Germany*

Abstract. Multi-particle beam dynamics calculations in presence of large beam currents have been carried out for a heavy ion Drift Tube Linac (DTL), in the framework of a European study group on Heavy Ion Driven Inertial Fusion (HIDIF). Linac design parameters were determined for high transmission and low emittance growth; then statistical errors as well as on-axis mismatch were added. The influence of field errors and different mismatch combinations on beam halo formation and emittance increase has been studied numerically, e.g. phase and amplitude jitters of the rf field, small changes of quadrupole gradients, mismatch of beam bunches at linac input. For proper ring injection, a transfer line and a bunch rotation cavity have to be inserted between linac and storage rings. The energy spread reduction after bunch rotation has been investigated both numerically and analytically, comparing an ideal case with a more realistic one which includes rf errors and mismatch.

THE HIDIF DRIVER SCENARIO

For 20 years concepts of possible energy production by heavy ion driven inertial confinement fusion have been investigated, a first phase ending with the HIBALL study in 1984 (1). The HIBALL driver scenario had already foreseen a combination of rf linacs and storage rings, with the fusion pellet hit by a couple of beams for direct ignition.

In the following years the target design has been changed: the beam energy is converted into radiation, which is supposed to ignite the pellet more symmetrically and to give a stable burn of the fuel. In this context a European HIDIF study group started to work on an appropriate new concept; the first goal was the conceptual design of an ignition facility to allow a proof of principle heavy ion fusion experiment; the advantage is the handling of much lower total beam energy at low repetition rates in comparison to a power plant for energy production.

The driver layout is still similar to the HIBALL scenario: the beams from 16 ion sources are pre-accelerated and merged together by several funnelling steps into one high-current beam, at each step doubling the current and the frequency. The main linac

---

[1] Work supported by BMBF and GSI-Darmstadt.

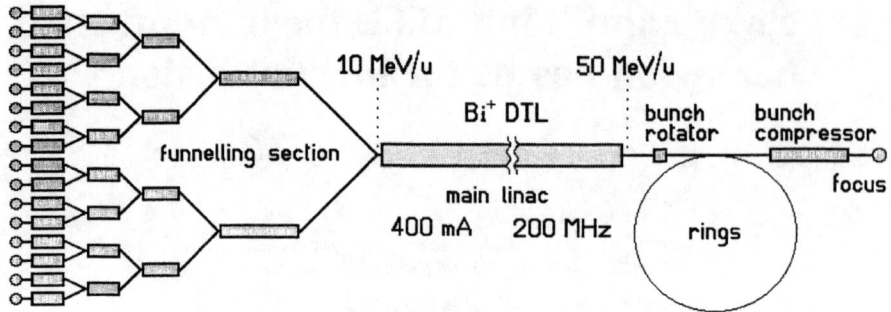

FIGURE 1. Scheme of the HIDIF driver scenario.

has to deliver a beam current of 400 mA ^{209}Bi$^+$ ions at an energy of 10 GeV. The beam is then accumulated and pre-compressed in some storage rings; finally it is further compressed in some induction linacs and focused onto the pellet (Fig. 1).

In this "tree-like" linac arrangement the ion sources and the low energy parts can be operated at necessarily moderate current levels, which facilitates to overcome the high space charge forces, but on the expense of more complexity. High current ion sources, transport, acceleration, chopping and funnelling of intense ion beams have been the subjects of several studies, compiled in (2).

In this paper the beam dynamics layout of the main linac will be investigated, including field errors and mismatch effects.

PARAMETER CHOICE OF THE MAIN LINAC

The beam dynamics layout of the linac is mainly orientated on the constraints imposed by the loss free injection into the following storage rings. The maximum transverse and longitudinal emittances, which still are accepted by the rings without losses, were taken as upper limits and tracked back to the linac input adding some safety margin for emittance growth along the linac.

By this procedure the maximum allowable size of the input emittances is determined, but not the particle distribution. In our calculations the common approach of a 4d waterbag distribution transversally, and a 2d waterbag longitudinally was chosen (Fig. 2-left). This choice cannot represent the real output distribution of the funnel tree, which is not known, since halo development and filamentation can be expected, caused by the multiple acceleration and beam merging steps before. As a first step in the framework of a feasibility study for an ignition facility, the 4d-2d waterbag distribution is fully sufficient to obtain substantial information on critical points, risks and possible cures.

For the main linac a conventional Alvarez type DTL operated at 200 MHz has been assumed, whose parameters are shown in Table 1. A 5F5D focusing scheme was adapted from a former study of a driver injector for heavy ion fusion (3); this leads to a smooth focusing ($\beta_{max} = 1.5\, \beta_{min}$) and limits the maximum pole tip magnetic field to

1.15 T for a bore radius of 16 mm in the beginning of the linac. The product of quadrupole length and gradient was kept constant along the linac. The electric field amplitude of 3.0 MV/m gives a calculated shunt impedance of 26 MΩ/m, which is a conservative value.

TABLE 1. Linac and beam parameters.

Mass number	209 (Bi+)	
Frequency	200.0	MHz
Current	400	mA
Number of cells N_c	9775	
Total length (10-50 MeV/u)	3383	m
Min. aperture radius	16	mm
Max. pole tip field	1.15	Tesla
Min. / max. electric field amplitude E_oT	2.80 - 2.88	MV/m
Total energy gain	40.0	MeV/u
Peak beam power (incl. 62.5% chopping)	690	kW/m
Peak dissipated power	320	kW/m
Total peak power	1.0	MW/m
Average shunt impedance	26	MΩ/m
Input / output trans. rms norm. emittance	0.176 / 0.183	π mm mrad
Input / output long. rms. emittance	1.66 / 1.83	π ns keV/u
Full current transverse tune σ_t	17 - 28	deg/period
Zero current transverse tune σ_{to}	22 - 31	deg/period
Full current longitudinal tune σ_l	11 - 6	deg/period
Zero current longitudinal tune σ_{lo}	13 - 10	deg/period
Transverse / longitudinal temperature	0.68 / 1.32	keV
Rms bunch radius / length	3.0 / 6.4	mm

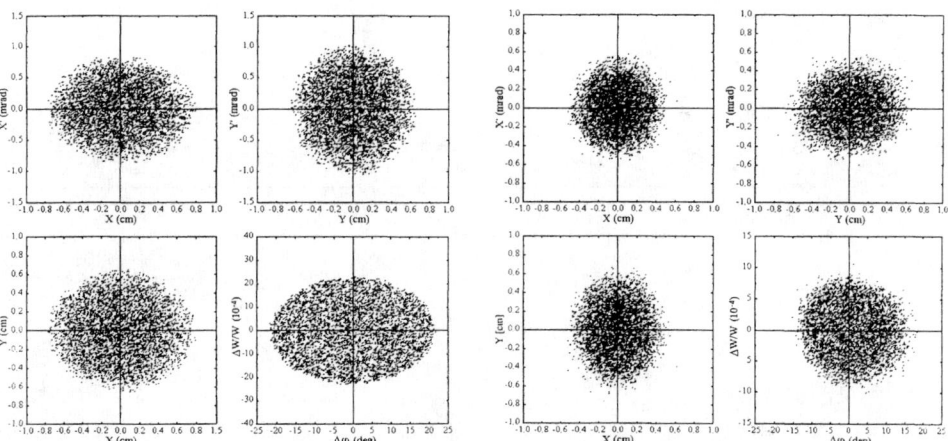

FIGURE 2. The 4d-2d waterbag input distribution for 5000 particles (left) and the corresponding output distribution at the end of the nominal linac (right).

The code CLAS was used to define the geometry of a few DTL cells, for a given β, and to compute the resulting electromagnetic fields in the gaps. The code GENLIN interpolates CLAS output for all the intermediate cells and generates the needed linac parameters. The code MAPRO uses GENLIN output to track thousands of macroparticles along the linac, with point-to-point space charge calculation (single particle solver): every half-cell, it computes the distance $d_{ij} = r_i - r_j$ for each pair of particles, the Coulomb force $F_{ij} = (q^2/4\pi\varepsilon_0) d_{ij} / |d_{ij}|^3$ and the new particle momentum $p_i + F_{ij} \Delta t$. A cut-off distance $d_{cut} = d_{ave}$ is introduced, to avoid that two particles come too close to each other, which would give unrealistic numerical effects: if $|d_{ij}| < d_{cut}$ then $|d_{ij}| = d_{cut}$.

For first quick calculations 1000 macroparticles were used, which leads to rather large fluctuations of the rms emittances along the linac. The calculations were repeated with 5000 and 20,000 macroparticles, to evaluate the influence of the particle number on the results (see Fig. 3). The rms beam behaviour becomes much smoother with increasing number of particles but no major changes could be observed in the size of the rms emittances. For a fast check 1000 particles are already sufficient; 5000 particles, as used in all following calculations, are a good compromise between computing time needed and loss of information.

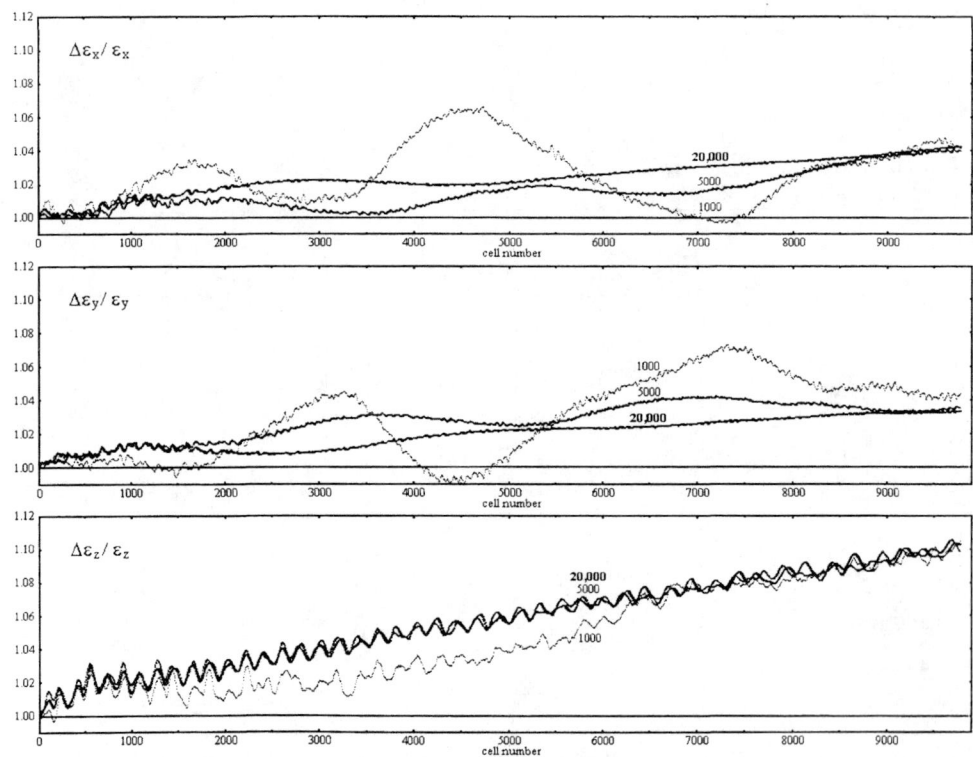

FIGURE 3. Behaviour of the rms emittance with 1000, 5000 and 20,000 particles along the linac.

In a final layout e.g. halo studies have to be done with much higher numbers of particles to guarantee particle losses as low as 10^{-5} to 10^{-6} per meter to avoid radio-activation of the structure.

In a first step calculations have been carried out to fix the design parameters (4,5). The input energy of the ions has to be 2 GeV to provide a sufficiently large acceptance for a 400 mA input beam. This energy and the frequency choice of 200 MHz are consistent with a four stage funnel tree, where in each step frequency and beam current are doubled. The first RFQ directly behind the ion source is then operated at 12.5 MHz and has to capture about 30 mA of Bi^+ ions, which is reasonable both for ion source and RFQ.

With the input emittances plotted in Figure 2-left, the beam behaviour along the linac was determined, resulting in the output emittances of Figure 2-right. The beam bunches are well confined, with small halo development only. For 99.5% of the particles the full transverse emittance is smaller than 4π mm mrad and the energy spread is smaller than 2×10^{-4} after debunching and bunch rotation. These values are in agreement with the requirements for ring injection. The beam radius along the linac is always smaller than 60% of the aperture radius, which can be considered as a good safety margin concerning particle losses. The linac length is about 3.3 km corresponding to 9775 cells.

In Figure 3 the behaviour of the rms emittances is plotted along the linac for three different numbers of macroparticles. The transverse rms emittance growth is less than 4% in both planes; the longitudinal rms emittance increases steadily along the linac up to about 10% at the end, which indicates that the chosen size of the input bunch is filling the linac bucket to quite a large extent.

THE EFFECT OF STATISTICAL ERRORS

Rf field amplitude and phase

In a next step, commonly used statistical errors of ±1% in amplitude and ±1° in phase (uncorrelated) have been added to the design values for the calculations. The same error has been attached to all gaps within a length of 2 m, which corresponds to an assumption that a 2 MW rf amplifier is needed to feed a 2 m long tank (see Tab. 1).

The results of beam dynamics calculations show that the effect of rf field errors is to displace the bunch centre with respect to the design phase and energy. But the bunch itself stays confined and moves as a whole, without any additional filamentation or emittance growth. An example of the bunch centre oscillations is shown in Figure 4 for an arbitrary set of errors.

The calculations have been repeated for 100 different sets of errors, finding that the position of the bunch centre at the linac end moves in phase space within an elliptic boundary with semi-axes of ±4° and ±3 MeV. To be precise, it follows a Gaussian distribution with variances 1.3° and 1 MeV; therefore the probability that the centre lies outside 3 times such variances at linac end is as small as 0.3%.

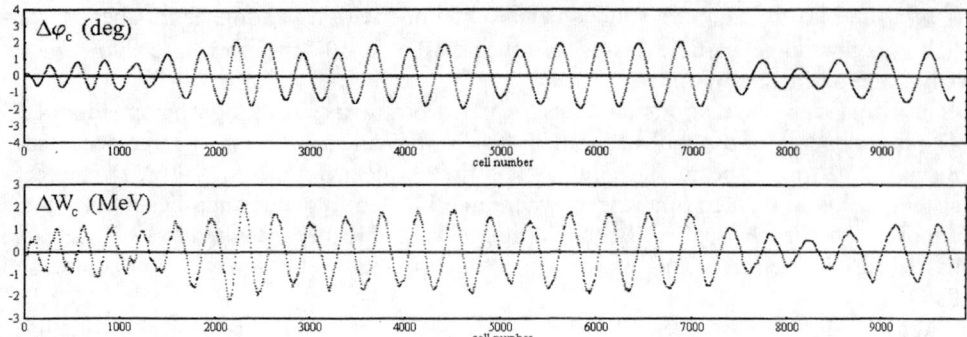

FIGURE 4. Bunch centre oscillations in phase and energy along the nominal linac, with 1% amplitude and 1° phase error in the rf field; 5000 particles.

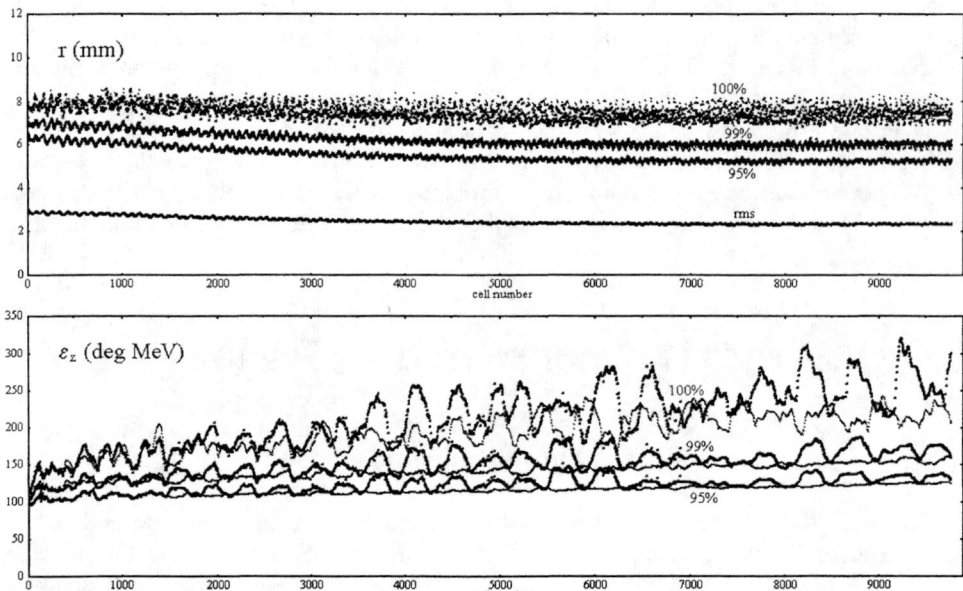

FIGURE 5. Development of the full (100%, 99% and 95%) radius and emittance along the nominal linac (thin line) and with 1% amplitude and 1° phase error in the rf field (bold line); 5000 particles.

Figure 5 shows the behaviour of the full beam radius and longitudinal emittance for 100%, 99% and 95% of the beam along the linac for the nominal case as well as in presence of rf field errors.

In the transverse planes, almost no effects can be observed with respect to the nominal case; the full beam radius always stays below 9 mm, compared to a 16 mm aperture.

In the longitudinal plane, the emittance is calculated by MAPRO assuming that the beam ellipse is always centred on the nominal energy and phase; therefore there is a "virtual" emittance increase if the oscillation of the bunch centre is not subtracted. But of course, as long as the bunch centre movement is not corrected, a larger area of longitudinal phase space at the linac end may be occupied by the shifting output beam, which has to be kept into account for the layout of the bunch rotation system.

Quadrupole gradients

At first a rather moderate statistical error of ±1% in field gradient was assumed. Since one power supply will feed the five quadrupoles of a kind in a period, the same error is assumed for all of them, which means a set of $N_c/5 = 1955$ different errors is used.

Without any correction, this resulted in a large emittance growth and halo formation, because the ±1% errors are statistically adding up. In order to cope with such an error, one has to rematch the beam. If this is not possible, then the error size has to be reduced to ±0.2%.

Figure 6 shows the full beam radius and longitudinal emittance for 100%, 99% and 95% of the beam, with ±1% and ±0.2% quadrupole gradient errors without rematching. Note that, with ±1% errors, the full beam radius approaches the pipe aperture and the full longitudinal emittance is also 60% larger than with ±0.2% errors, as for high intensity bunched beams a transverse mismatch will produce longitudinal halo, as described in next section.

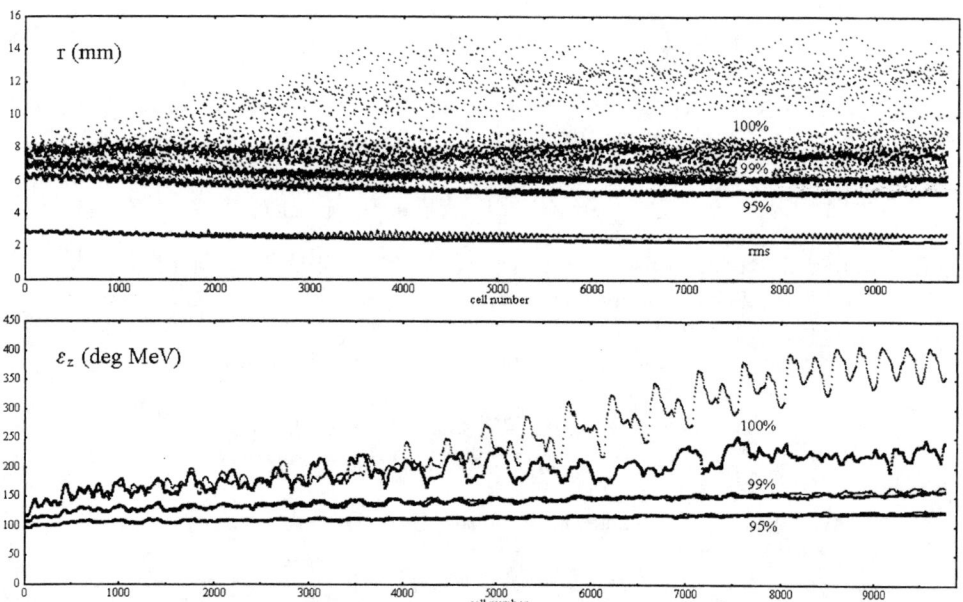

FIGURE 6. Development of the full (100%, 99% and 95%) radius and emittance with 5000 particles along the nominal linac, with quadrupole gradient errors of 1% (thin curve) and 0.2% (bold curve).

MISMATCH OF BUNCHED BEAMS

For a <u>matched dc</u> beam, with identical emittances and tunes in both transverse planes, the rms beam radius R is constant along a continuously focusing channel. Assuming a small mismatch and linearising it, the mean radius performs small oscillations of amplitude $x_0, y_0 \ll R$ around the average value. These oscillations can be described in terms of the superposition of two "eigenmodes": one in which the amplitudes in the two perpendicular transverse directions x and y are 180° out-of-phase and one in which they are in-phase (6). The phase advances ϕ_1 and ϕ_2 of these two oscillation modes in one focusing period are:

$$\phi_1 = (\sigma_0^2 + 3\sigma^2)^{1/2} \quad \text{and} \quad \phi_2 = (2\sigma_0^2 + 2\sigma^2)^{1/2} \qquad (1)$$

where σ_0 and σ is the tune-shift for zero and full current; for zero current ($\sigma_0 = \sigma$), one has $\phi_1 = \phi_2 = 2\sigma_0$.

For a <u>mismatched bunched</u> beam with identical emittances and tunes in both transverse planes, it is possible to describe the beam envelope by means of 3 "eigenmodes" of oscillation, whose features can be approximated by analytical functions of the four tunes σ_l, σ_t, σ_{lo}, σ_{to} (longitudinal, transverse, with and without current) (7).

The so-called <u>quadrupolar</u> mode is excited by increasing the horizontal bunch size a_x and simultaneously decreasing the vertical bunch size a_y by the same amount, while the bunch length b is not changed $(+ - 0)$, for example:

$$\Delta a_x/a_x = -\Delta a_y/a_y = +20\% \quad \text{and} \quad \Delta b/b = 0 \qquad (2)$$

The mismatches in x and y will oscillate in phase opposition, while the z size will stay constant (Fig. 7). The phase advance is given by:

$$\sigma_Q = 2\,\sigma_t \qquad (3)$$

The numerical calculations came up with $\sigma_Q = 38°$, which is in fact larger than the theoretical $2\,\sigma_t = 34°$ but smaller than the phase advance of $\phi_1 = 39.3°$ obtained for the out-of-phase mode in a dc beam.

In the <u>high</u> mode, a_x, a_y and b are all increased $(+ + +)$, according to:

$$\Delta a_x/a_x = \Delta a_y/a_y = +20\% \quad \text{and} \quad \Delta b/b = f_H\, \Delta a_x/a_x \qquad (4)$$
$$f_H = [\sigma_H^2 - 2(\sigma_{to}^2 + \sigma_t^2)] / (\sigma_{to}^2 - \sigma_t^2) \qquad (5)$$

The mismatches in x, y and z are all in phase. The phase advance is given by:

$$\sigma_H^2 = A + B^{1/2} \qquad (6)$$
$$A = \sigma_{to}^2 + \sigma_t^2 + \tfrac{1}{2}(\sigma_{lo}^2 + 3\sigma_l^2) \qquad (7)$$
$$B = [\sigma_{to}^2 + \sigma_t^2 - \tfrac{1}{2}(\sigma_{lo}^2 + 3\sigma_l^2)]^2 + 2(\sigma_{to}^2 - \sigma_t^2)(\sigma_{lo}^2 - \sigma_l^2) \qquad (8)$$

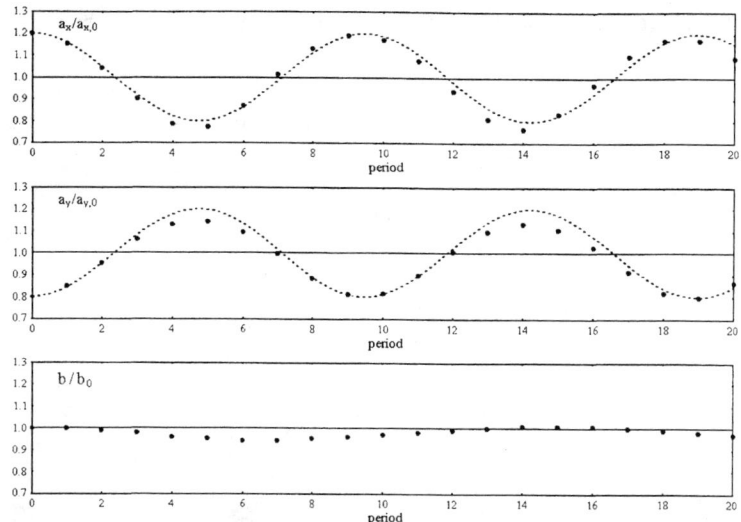

FIGURE 7. Excitation of the quadrupolar mode through 20% initial mismatch: rms bunch sizes with respect to the nominal ones at the end of each period, for the 4d-2d waterbag input distribution, 5000 particles. The dotted line is the best fit.

One has $f_H = 0.093$, therefore $\Delta b/b = 1.8\%$. Numerical calculations showed that the initial theoretical mismatch values of $\Delta a_x/a_x = \Delta a_y/a_y = +20\%$ and $\Delta b/b = +1.8\%$ indeed are exciting the pure high mode oscillation, with a calculated phase advance $\sigma_H = 41°$, very close to the theoretical value of $39.5°$.

In the <u>low</u> mode, a_x and a_y are decreased, while b is increased $(--+)$, according to:

$$\Delta b/b = +20\% \quad \text{and} \quad \Delta a_x/a_x = \Delta a_y/a_y = (1/f_L) \Delta a_x/a_x \tag{9}$$
$$f_L = [\sigma^2_L - 2(\sigma^2_{to} + \sigma^2_t)] / (\sigma^2_{to} - \sigma^2_t) \tag{10}$$

The mismatches in x and y are in phase, while the mismatch in z is of opposite phase. The phase advance is given by:

$$\sigma_L^2 = A - B^{1/2} \tag{11}$$

One has $f_L = -5.29$, therefore $\Delta a_x/a_x = \Delta a_y/a_y = -3.8\%$. Numerical calculations showed that the initial theoretical mismatch values of $\Delta a_x/a_x = \Delta a_y/a_y = -3.8\%$ and $\Delta b/b = +20\%$ indeed are exciting the pure low mode oscillation, with a calculated phase advance $\sigma_L = 25°$, very close to the theoretical value of $22.7°$.

The advantage of this model is that, for small initial mismatch amplitudes of the beam, the envelope is a superposition of the three above described eigenmodes.

If either the beam centre is displaced transversally due to misalignments or longitudinally due to rf amplitude and phase errors, all above given statements are correct for the oscillation of the beam envelope relative to the shifted bunch centre.

Halo formation due to mismatch for the HIDIF linac

For the quadrupolar mode (+ − 0) there is an immediate radial halo formation, but no axial one. The radial rms emittance is increasing, since the local aspect ratio is higher in the mismatched case:

$$a_x(s) / a_y(s) = 1.44 \, a_{x,o}(s) / a_{y,o}(s) \qquad (12)$$

For the high mode (+ + +) there is a delayed radial halo formation, but no axial one. The radial rms emittance is only slightly increasing, since the aspect ratio is not changed by the mismatch:

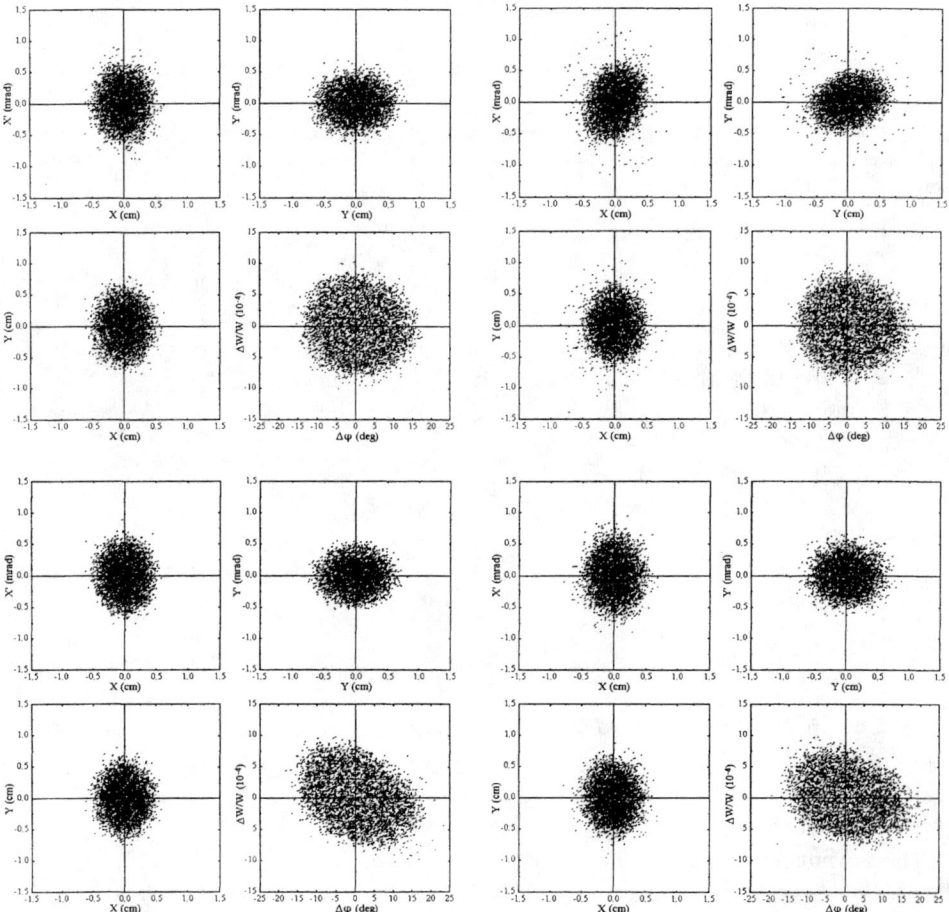

FIGURE 8. Output distribution of 5000 particles with maximum 20% initial mismatch error. *Top left*: quadrupolar mode. *Top right*: high mode. *Bottom left*: low mode. *Bottom right*: mixed mode.

$$a_x(s) / a_y(s) = a_{x,o}(s) / a_{y,o}(s) \qquad (13)$$

For the <u>low</u> mode (− − +) there is a delayed <u>axial halo</u> formation, but no radial one, since the axial rms emittance increase is small.

A "<u>mixed</u>" mode ($\Delta a_x/a_x = -\Delta a_y/a_y = \Delta b/b = +20\%$) is the worst case, leading to both radial and axial halo formation.

Figure 8 shows the output distribution of 5000 particles with a maximum 20% mismatch error in the three "pure" modes and in the "mixed" one.

The observed either radial or axial halo formation when exciting one of the pure bunched beam eigenmodes can be explained in terms of 1/2 parametric resonance excitation between single particle and oscillating mismatched core (8,9). Due to the non-linear part of the space charge forces, the oscillation frequency of single particles is limited between the full and the zero current tune either in the transverse or longitudinal plane, which are different at the beginning of the HIDIF linac (see Tab. 1). For the above given frequencies of the three beam eigenmodes, the excitation of one of them at the beginning of the linac will lead either to radial or axial halo formation only. The mixed mode, which is a superposition of the quadrupolar and of the low eigenmodes, leads simultaneously to radial and axial halo formation.

In summary, the four mismatch cases considered here give rise to some emittance degradation but they do not blow up the beam over the limits of ring injection, as long as the mismatch is limited to a maximum of 20%.

TRANSFER LINE AND BUNCH ROTATOR

The most severe requirement for loss-free injection into the rings is a momentum spread $dp/p < \pm 0.02\%$. The output momentum spread in the nominal linac is more than double; moreover, the bunch centre can be displaced by another $\pm 0.02\%$ due to rf field errors.

A 170 m long transfer line was designed, keeping the same 5F5D focusing scheme of the main linac, with constant focusing and cell length. At the end of this line, a bunching system will rotate the bunch.

The beam behaviour in this transfer line has been examined with MAPRO code, using the nominal linac (no errors) output distribution for 5000 particles as an input. The total bunch length increases from $\pm 15°$ to $\pm 85°$ and, due to space charge forces, the energy spread increases from $\pm 0.09\%$ to $\pm 0.15\%$ (see Fig. 9-left), while the radial dimensions stay constant. The momentum spread is then reduced by an ideal linear bunch rotation cavity. Figure 10-left shows that the momentum spread is well below the requirements, after rotation.

For a more realistic simulation, the output of the linac in case of the "mixed" mode mismatch was used, displacing the bunch centre in order to account for rf field errors. From numerical simulations (10) we know that, with 99.7% probability, the bunch centre offset at the linac end will lie within an ellipse with $\Delta\phi_{c,0}^{max} = \pm 4°$ and $\Delta W_{c,0}^{max} = \pm 3$ MeV; then, as an example, a combination of phase and energy shift on the elliptic

TABLE 2. Transfer line parameters.

Bunch length at input b_0	±15	deg
Energy spread at input	±0.09	%
Bunch length at output b_L	±85	deg
Energy spread at output	±0.15	%
Total length	170	m
1st harmonic amplitude U_1	14.2	MV
3rd harmonic amplitude U_3	1.58	MV
5th harmonic amplitude U_5	0.57	MV
Bunch length after rotation	±85	deg
Energy spread after rotation	±0.02	%

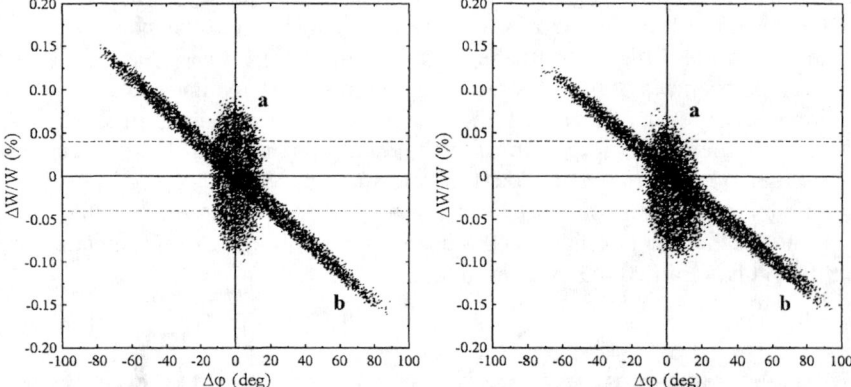

FIGURE 9. Beam dynamics calculations: output distribution of 5000 particles: (a) at the end of the nominal linac, (b) at the end of the transfer line before rotation. *Left*: nominal layout. *Right*: 20% "mixed mode" mismatch, bunch centre shifted by +3° and −2 MeV (due to rf errors).

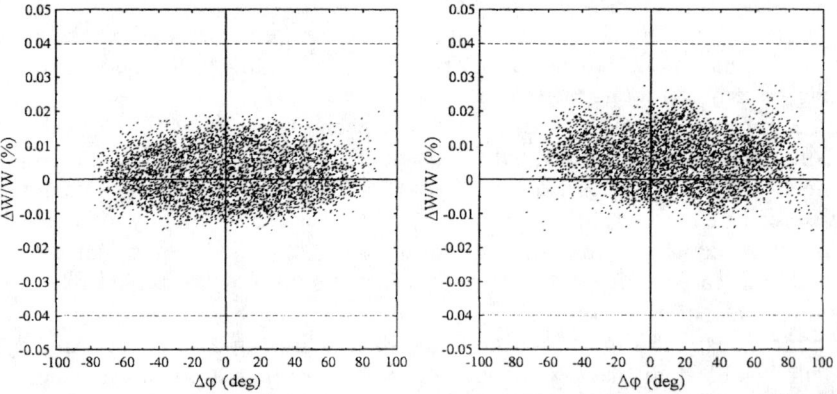

FIGURE 10. Beam dynamics calculations: output distribution of 5000 particles at the end of the transfer line after rotation. *Left*: nominal layout and linear rotation. *Right*: 20% "mixed mode" mismatch, bunch centre shifted by +3° and −2 MeV and 3-harmonic rotation system.

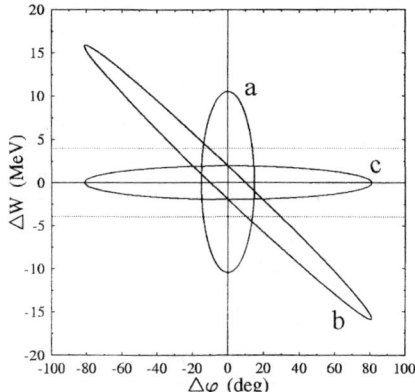 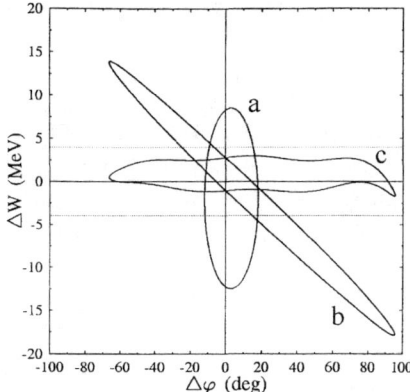

FIGURE 11. Analytical approach: output ellipses: (a) at the end of the linac, (b) at the end of the transfer line before rotation and (c) after rotation. *Left*: nominal layout and linear rotation. *Right*: bunch centre at linac end shifted by +3° and −2 MeV, and 3-harmonic bunch rotation.

boundary was considered: $\Delta\phi_{c,0} = 3°$, $\Delta W_{c,0} = -2$ MeV. Moreover, a 3-harmonic bunch rotation system was used. Results from beam dynamics calculations are shown in Figure 9-right and Figure 10-right.

To evaluate analytically the rf error acceptance, at first the beam ellipse in the longitudinal plane was plotted at the end of the nominal linac, before and after linear bunch rotation (see Figure 11-left), using the Twiss parameters and 9 times the rms emittance obtained from numerical simulations.

Then, in case of rf errors in the DTL, the bunch centre offset was superposed to the longitudinal beam ellipse at the end of the linac and transported along the transfer line. Since the bunch centre behaves like a single particle with zero current, the energy offsets ΔW_c does not change, while the phase offset $\Delta\phi_c$ is given by:

$$\Delta\phi_{c,L} = \Delta\phi_{c,0} + (360°/\lambda\,(\beta\gamma)^3\,m_0c^2)\,L\,\Delta W_{c,0} = \Delta\phi_{c,0} + (5.8°/\text{MeV})\,\Delta W_{c,0} \quad (14)$$

where the subscripts 0 and L indicate respectively the end of the linac and of the transfer line. The bunch rotator will finally change the energy offset according to:

$$\Delta W_{c,L} = \Delta W_{c,0} - eU \sin(\Delta\phi_{c,L}) \quad (15)$$

The longitudinal phase space ellipses at the end of the linac and of the transfer line, before and after a 3-harmonic bunch rotation, are shown in Figure 11-right; the Twiss parameters and rms emittance were obtained from the numerical results shown in Figure 9-left.

We could show that, even including rf errors, all particles are in the requested range after bunch rotation: the ellipse always stays between ±4 MeV, as requested for loss-free ring injection. This means that the bunch rotation system itself is able to correct an initial displacement of the bunch centre.

INJECTOR LINAC FOR A HIDIF POWER PLANT

For a HIDIF power plant on contrary to an ignition facility the average linac power will be as high as 150 MW at high repetition rates. One key issue of such high intensity accelerator facilities is to avoid particle losses in the linac as well as at ring injection, i.e. to reduce halo development and axial filamentation of the beam. For "hands on maintenance" in the linac a deposited beam power of less than 1.5 kW/m is required, which limits particle losses to 10^{-5} to 10^{-6}/m; the maximum energy deviation from the design value has to be smaller than ±4 MeV for loss free ring injection, including possible bunch centre offsets due to rf errors.

Up to now calculations, also with 20,000 particles, ended with no transverse losses in the linac and energy spreads below the upper limit, including mismatch in case of an unfilamented 4d-2d waterbag distribution at the linac input (11). For a realistic filamented 6d input, which is not yet known, additional safety margins must be foreseen. Since the assumed radial and axial emittances are already small and should not be reduced further, the acceptance of the linac has to be increased. For this option some reserve is left e.g. larger bore radii, increase of the synchronous phase or a higher electric field strength in modern DTL structures.

As shown in Figure 11, there is some safety margin left, to have no particles outside ±4 MeV after bunch rotation, even increasing the full emittance from 9 to 16 times the rms one and including bunch centre shifted by 3° and −2 MeV.

Further calculations are underway to study the possibility of the "telescoping option" for the final beam transport system (2): three different heavy ion species with mass differences of up to 20% have to be accelerated to the same final momentum, i.e. different ion energies in the linac. Only the lightest ions are accelerated to the linac end; the heavier ones have to be transported through rather large parts of the linac without loss of beam quality, which up to now seems not to be possible (11).

CONCLUSIONS

In the framework of a study for a Heavy Ion Driven Ignition Facility (HIDIF) the particle dynamics layout for a heavy ion DTL have been investigated to allow for the acceleration of a 400 mA Bi^+ beam to an energy of 10 GeV with high transmission and low emittance increase. The required beam quality is defined by the demand for loss free injection into the following accumulator rings: transverse emittances of less than 4π mm mrad and a maximum momentum spread of less than 2×10^{-4}.

A parameter set could be found for the linac, which is rather conventional: frequency 200 MHz, E_oT = 2.8 - 2.9 MV/m, synchronous phase 40° - 30°, maximum quadrupole pole tip field of about 1 T. Due to the low charge state, the linac length sums up to 3.3 km for a total energy gain of 8 GeV. Multi-particle simulations with up to 20,000 macroparticles have been performed to check the parameter choice with a standard 4d transversal and 2d longitudinal waterbag distribution. The size of the input emittances was determined by tracking back the maximum allowable linac output emittances including some small safety margin for emittance growth.

With transverse input emittances of 0.176 π mm mrad and longitudinal one of 1.66 π ns keV/u, only small increases of the rms values were seen in the error-free case; no major beam halos developed. Adding moderate statistical errors in rf field phase and amplitude resulted in bunch centre oscillations, the bunch itself staying well confined in all three phase space planes. Quadrupole gradient errors must be kept small or corrected carefully because their influence sums up, due to the huge number of quadrupoles along the linac.

Different cases of initial mismatch for on-axis beams were studied; the analytically predicted oscillation modes could be identified in the numerical calculations. For two modes some longitudinal filamentation was found, that is tolerable as long as the mismatch factor is smaller than 20% in all phase space planes. The transverse beam behaviour turned out to be less critical, only up to 2/3 of the aperture were filled by the beam.

A preliminary layout of a beam transfer line from linac to the rings was made, including a bunch rotation system. After debunching of the beam, with some increase in the energy spread due to the space charge forces, in all cases the bunch could be rotated for ring injection below the critical energy spread of ±4 MeV. Moreover the bunch rotation proved to be self-regulating for not too big bunch centre offsets at the linac output.

The solution found turned out to be rather stable against errors and mismatch, which may be due to the fact that the particle dynamics layout is not space charge dominated.

REFERENCES

1. R. Badger et al., "HIBALL - A Conceptual Heavy Ion Beam Driven Fusion Reactor Study", KfK 3203, Karlsruhe and UWFDM-450 (1981).
2. Hofmann, I., and Plass, G., Ed., "The HIDIF Study – Report of the European Study Group on Heavy Ion Driven Inertial Fusion for the period 1995-1998", GSI-98-06 REPORT, Darmstadt, August 1998.
3. Koshkarev, D.G., Korenev, I.L., and Yudin, L.A., "Conceptual design of linac for power HIF driver", in *Proceedings of the 18th International Linac Conference*, Geneva, Switzerland, August 1996, pp. 423-425.
4. Parisi, G., and Deitinghoff, H., "Considerations on Particle Dynamics in a Heavy Ion DTL", Proc. of PAC '97, Vancouver, B.C., 12-15 May 1997, p 1900.
5. Parisi, G., Deitinghoff, H., Bongardt, K., and Pabst, M., "A heavy ion DTL design for HIDIF", 12th Int. Symp. on Heavy Ion Inertial Fusion, HIIF 97, Heidelberg, Germany, September 1997, *Nuclear Instr. and Meth. in Phys. Res. Sect. A* (415)1-2 (1998) p. 332.
6. Struckmeier, J., and Reiser, M., "Theoretical Studies of Envelope Oscillations and Instabilities of Mismatched Intense Charged Particle Beams in Periodic Focusing Channels", GSI-83-11, April 1983.
7. Pabst, M., and Bongardt, K., "Analytical approximation of the three mismatch modes for bunched beams", ESS 97-85-L, Jülich, August 1997.
8. Pabst, M., Bongardt, K., and Letchford, A., "Progress on intense proton beam dynamics and halo formation", in *Proceedings of the 6th European Particle Accelerator Conference*, Stockholm, Sweden, June 1998, pp. 146-150.
9. Bongardt, K., Pabst, M., and Letchford, A., "Halo Formation by Mismatch for the High Intensity Bunched Beams", in *Proceedings of the 19th International Linac Conference*, Chicago, IL, August 1998, to be published (see http://www.aps.anl.gov/conferences/LINAC98/papers/TH4025.pdf).
10. Parisi, G., Deitinghoff, H., Bongardt, K., and Pabst, M., "Error Effects and Parameter Analysis for a HIDIF DTL", in *Proceedings of the 6th European Particle Accelerator Conference*, Stockholm, Sweden, June 1998, pp. 1121-1123.
11. Parisi, G., Ph.D. thesis, in progress.

Halo Formation in Anisotropic Beams

Masanori Ikegami

Proton Accelerator Laboratory, Japan Atomic Energy Research Institute
Tokai-mura, Naka-gun, Ibaraki-ken, 319-1195 JAPAN
E-mail: ikegami@linac.tokai.jaeri.go.jp

Abstract. We have applied the particle-core analysis to transversely isotropic and anisotropic beams executing general envelope oscillation which is a superposition of two normal modes. We find that, in isotropic cases with the same emittance and external focusing strength in the two transverse directions, single particle motion exhibits strong chaosity in wide parameter space due to the resonance overlap of two 2:1 particle-core resonances. The strong chaosity is expected to result in an increase of halo intensity enhancing the chance for the particles to gain excess energy through resonant interaction with the core. On the other hand, it is found that the resonance overlap can be avoided by choosing appropriate parameters in anisotropic situations. The avoidance of the resonance overlap suppresses chaosity in single particle motion, which strongly suggests that halo intensity can be reduced in anisotropic situations by choosing appropriate parameters.

INTRODUCTION

Space-charge dominated beams are known to exhibit various nonlinear behaviors such as charge-redistribution, structure resonance, equipartitioning and beam halo formation. To achieve high-intensity beams required for a wide variety of applications, it is essential to understand these phenomena in both qualitative and quantitative ways. From this point of view, space-charge induced phenomena have been studied for several decades. Especially, halo formation mechanism has been studied extensively in the past several years, because it turns out to be essential to keep the beam loss rate extremely low in realizing high-intensity ion accelerators now proposed for a spallation neutron source, transmutation of nuclear waste, heavy ion fusion, *etc.* In these studies, the so-called *particle-core model* (1,2) has frequently been adopted. In this model, we consider a beam core oscillating due to initial beam-size mismatch and test particles initially located outside the core, and the stability properties of test particles are examined assuming that the core oscillation is not affected by the test particle motion.

In a number of numerical analyses based on this model, it has so far been found that the 2:1 parametric resonance between test particle oscillation and breathing core oscillation is a main cause of halo formation (1). However, in most analyses, a core executing pure breathing oscillation is assumed in contrast to the fact that general core oscillation is a superposition of the breathing and quadrupole modes of oscillation. The simultaneous excitation of these two modes is expected to cause stronger chaos in test particle motion due to resonance overlap and result in an increase of halo intensity. However, very few attempts have so far been made to confirm this expectation directly. In addition, in most preceding works, only isotropic cases are considered, where the emittance and external focusing force strength are the same in the two transverse directions. It is obviously insufficient recalling that beam halos may also be formed in a circular machine and cause a serious beam loss.

In this paper, we examine the effects of anisotropy of beam cores on beam halo formation. At the same time, we also deal with the effects of a superposition of two normal modes on halo dynamics because these two problems are closely related as discussed later. In the next section, we first present the numerical method to apply the particle-core model to anisotropic beams. Then, we show some typical results obtained by applying the method to isotropic and anisotropic beams. After discussions, a summary is given in the last section. In the present study, we will restrict our treatment to test particles with zero angular momentum and coasting beam cores with the Kapchinskij-Vladimirskij, or KV, distribution (3). The external focusing field in an actual machine is periodic, but here we adopt the smooth approximation (4) for simplicity.

NUMERICAL METHOD

Core oscillation

The time-evolution of core oscillation is governed by the envelope equations. In terms of scaled variables, the envelope equations can be written as

$$\frac{d^2 a}{d\tau^2} = -a + \frac{2K}{a+b} + \frac{1}{a^3}, \tag{1}$$

and

$$\frac{d^2 b}{d\tau^2} = -\zeta^2 b + \frac{2K}{a+b} + \frac{\eta^2}{b^3}, \tag{2}$$

where a and b are respectively the scaled core width in the horizontal and vertical direction, and the independent variable τ is the scaled distance along the beam line. The scaled space-charge perveance K is a measure of beam density. The parameter ζ is the horizontal and vertical external focusing strength ratio, and η is the horizontal and vertical emittance ratio. Matched envelope a_0 and b_0 can be obtained by solving the following equations,

$$a_0 = \frac{2K}{a_0 + b_0} + \frac{1}{a_0^3}, \quad (3)$$

and

$$\zeta^2 b_0 = \frac{2K}{a_0 + b_0} + \frac{\eta^2}{b_0^3}. \quad (4)$$

Here, we approximate the core oscillation with small mismatch as

$$a = a_0[1 + \delta a \exp(iks)], \quad (5)$$

and

$$b = b_0[1 + \delta b \exp(iks)]. \quad (6)$$

Substituting Eqs. (5) and (6) into Eq. (1) and dropping second or higher order terms of δa and δb, we find

$$k^2 = 1 + \frac{2K}{(a_0 + b_0)^2}\left(1 + \frac{\delta b}{\delta a}\right) + \frac{3}{a_0^4}. \quad (7)$$

In an analogous way, we find from Eqs. (5), (6) and (2)

$$k^2 = \zeta^2 + \frac{2K}{(a_0 + b_0)^2}\left(1 + \frac{\delta a}{\delta b}\right) + \frac{3\eta^2}{b_0^4}. \quad (8)$$

Solving Eqs. (7) and (8) simultaneously, we can obtain the normal mode frequency k and the relative horizontal and vertical amplitude of a normal mode $\delta a/\delta b$. Because Eqs. (7) and (8) have two solutions for k, two modes are present, namely; the breathing (high-frequency) mode k_H and the quadrupole (low-frequency) mode k_L. The horizontal and vertical oscillations are in phase in the breathing mode, while those are 180° degree out of phase in the quadrupole mode.

We here introduce tune depression in the horizontal direction μ_x as another measure of beam density. The parameter μ_x is defined as the ratio of the space-charge depressed phase advance to the zero-current phase advance for the horizontal motion. Furthermore, we introduce aspect ratio of the matched beam $\chi = b_0/a_0$ as another measure of anisotropy.

In terms of μ_x, χ and ζ, the normal mode oscillation frequency is written as

$$k_H^2 = 2(1+\zeta^2) - (1-\mu_x^2)\frac{3+4\chi+3\chi^2}{2\chi(1+\chi)}$$
$$+\sqrt{\left[2(1-\zeta^2)+(1-\mu_x^2)\frac{3(1-\chi)}{2\chi}\right]^2 + \frac{(1-\mu_x^2)^2}{(1+\chi)^2}},\quad (9)$$

for the breathing mode and

$$k_L^2 = 2(1+\zeta^2) - (1-\mu_x^2)\frac{3+4\chi+3\chi^2}{2\chi(1+\chi)}$$
$$-\sqrt{\left[2(1-\zeta^2)+(1-\mu_x^2)\frac{3(1-\chi)}{2\chi}\right]^2 + \frac{(1-\mu_x^2)^2}{(1+\chi)^2}},\quad (10)$$

for the quadrupole mode. The relative horizontal and vertical oscillation amplitude of normal modes can be obtained as

$$\frac{\delta a_H}{\delta b_H} = (1+\chi)\frac{k_H^2 - \zeta^2\left[4 - \frac{1}{\zeta^2\chi}(1-\mu_x^2)\left(3 - \frac{\chi}{1+\chi}\right)\right]}{1-\mu_x^2}, \quad (11)$$

for the breathing mode and

$$\frac{\delta a_L}{\delta b_L} = (1+\chi)\frac{k_L^2 - \zeta^2\left[4 - \frac{1}{\zeta^2\chi}(1-\mu_x^2)\left(3 - \frac{\chi}{1+\chi}\right)\right]}{1-\mu_x^2}, \quad (12)$$

for the quadrupole mode, where δa_H and δb_H are the horizontal and vertical oscillation amplitude for the breathing mode, and δa_L and δb_L are those for the quadrupole mode.

As the general core oscillation is a superposition of the breathing and quadrupole modes, the core oscillation with small mismatch can be approximated as

$$a = a_0\left[1 + \delta a_H \cos(k_H \tau) + \delta a_L \cos(k_L \tau)\right], \quad (13)$$

and

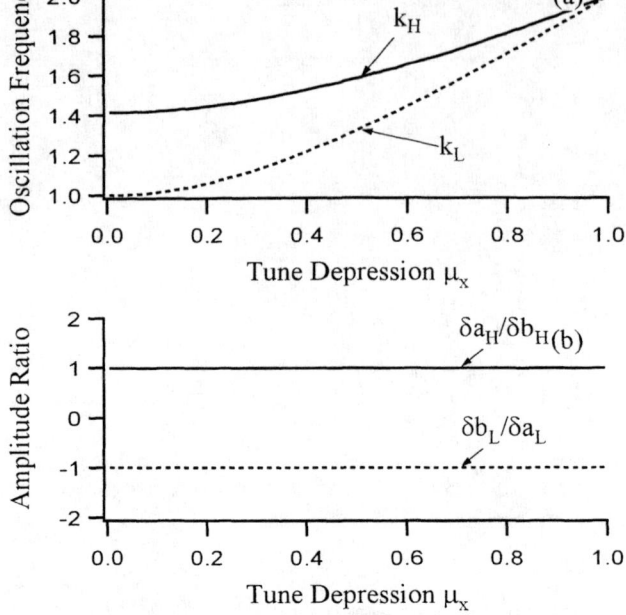

FIGURE 1. Beam density dependence of frequency and relative horizontal and vertical oscillation amplitude of normal modes (isotropic case). (a) Normal mode frequency. (b) Relative horizontal and vertical oscillation amplitude.

$$b = b_0 \left[1 + \delta b_H \cos(k_H \tau) + \delta b_L \cos(k_L \tau) \right]. \tag{14}$$

For later reference, we here introduce mismatch factor M which is defined by

$$M = \sqrt{\frac{\delta a_H^2 + \zeta^2 \delta b_H^2 + \delta a_L^2 + \zeta^2 \delta b_L^2}{2}}. \tag{15}$$

In addition, we here introduce another parameter Γ which is defined by

$$\Gamma = \frac{\left(\delta a_H^2 + \zeta^2 \delta b_H^2\right) - \left(\delta a_L^2 + \zeta^2 \delta b_L^2\right)}{\left(\delta a_H^2 + \zeta^2 \delta b_H^2\right) + \left(\delta a_L^2 + \zeta^2 \delta b_L^2\right)}. \tag{16}$$

The mismatch factor M represents total strength of mismatch oscillation, while the parameter Γ represents relative strength of two normal modes, namely; $\Gamma=1$ for the pure breathing mode cases, $\Gamma=-1$ for the pure quadrupole mode cases, and $\Gamma=0$ for the cases

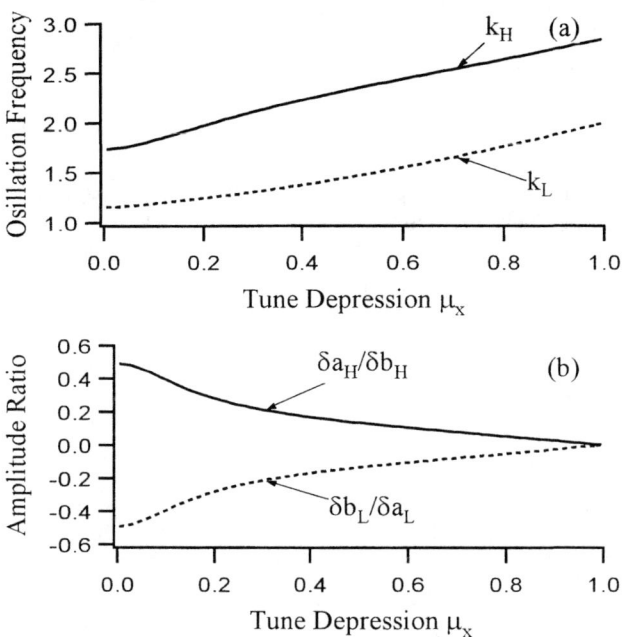

FIGURE 2. Beam density dependence of frequency and relative horizontal and vertical oscillation amplitude of normal modes (anisotropic case). The parameters are set to be $\zeta=1.42$ and $\eta=1$. (a) Normal mode frequency. (b) Relative horizontal and vertical oscillation amplitude.

where both two modes are equally excited.

Let us here show the parameter dependence of some characteristics of normal modes. Figures 1 shows the density dependence of frequency and relative horizontal and vertical oscillation amplitude of normal modes in isotropic cases. We readily see in Fig. 1 that k_H and k_L exist between one and two. It is also seen in Fig. 1 that the absolute value of the horizontal and vertical oscillation amplitude ratio is always one regardless of beam density. Figures 2 shows the density dependence of frequency and relative horizontal and vertical oscillation amplitude of normal modes in an anisotropic case. In this case, as readily seen in Fig. 2, k_L still exists between one and two while k_H moves upward. This frequency shift affects the test particle stability as discussed later. It is also seen in Fig. 2 that the absolute value of the horizontal and vertical oscillation amplitude ratio is not one and depends on beam density in anisotropic cases. We also see in Fig. 2 that, in this case, the breathing mode is primarily vertical ($|\delta a_H| < |\delta b_H|$) while the quadrupole mode is primarily horizontal ($|\delta a_L| > |\delta b_L|$). To be noted here is that the ratio $\delta a_H / \delta b_H$ gives the strength of coupling between the horizontal and vertical oscillation of the core. Then, we can clearly see in Fig. 2 that the coupling strength is dependent on beam density in anisotropic cases.

Test-particle equations of motion

As we assume that there is no coupling between the horizontal and vertical motion except for the space charge force, the beam ellipse remains to be upright in real space. In addition, we only consider the test particles with zero angular momentum. Therefore, the motion of the test particles initially located on the horizontal or vertical plane is restricted on that plane. Thus, the equation of motion for the test particle initially located on the horizontal plane ($y=0$ and $dy/d\tau=0$) can be written (5) as

$$\frac{d^2x}{d\tau^2} + x - \frac{2K}{a(a+b)}x = 0 \qquad (|x| \leq a), \qquad (17)$$

and

$$\frac{d^2x}{d\tau^2} + x - \frac{2K}{x^2 + |x|\sqrt{x^2+b^2-a^2}}x = 0 \qquad (|x| > a). \qquad (18)$$

In an analogous way, the equation of motion for the test particle initially located on the vertical plane ($x=0$ and $dx/d\tau=0$) can be written as

$$\frac{d^2y}{d\tau^2} + \zeta^2 y - \frac{2K}{b(a+b)}y = 0 \qquad (|y| \leq b), \qquad (19)$$

and

$$\frac{d^2y}{d\tau^2} + \zeta^2 y - \frac{2K}{y^2 + |y|\sqrt{y^2+a^2-b^2}}y = 0 \qquad (|y| > b). \qquad (20)$$

Numerically integrating Eqs. (17) to (20) with Eqs. (13) and (14), we obtain the time-evolution of test particle motion. In the integration, the fourth-order symplectic integration algorithm (6) is employed.

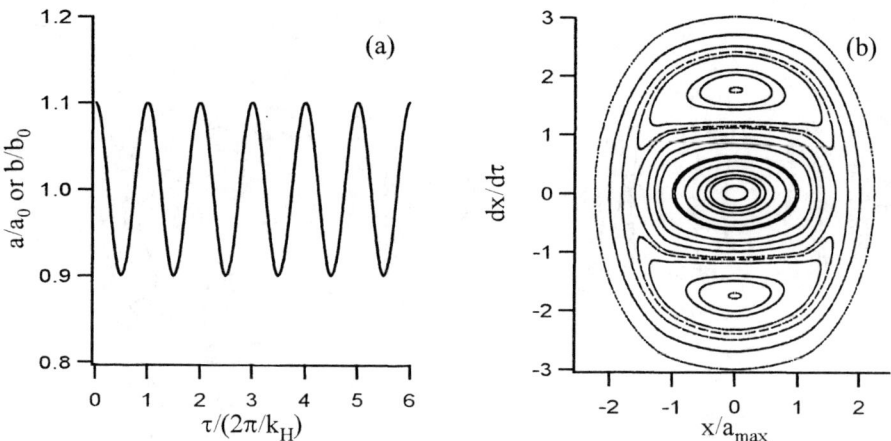

FIGURE 3. Results for an isotropic beam (pure breathing mode case). The parameters are set to be $\zeta=1$, $\eta=1$, $\mu_x=0.49$, $M=0.1$, and $\Gamma=1$. (a) Time-evolution of the beam core. The solid and broken lines respectively represents the horizontal and vertical core half-width scaled by the matched one. In this figure, these two lines lie one upon another. (b) Poincaré plots for test particles. The abscissa is the horizontal coordinate scaled by the maximum horizontal core width a_{max}.

NUMERICAL RESULTS

Isotropic Cases

First, we will consider isotropic cases where $\zeta=1$ and $\eta=1$. Figures 3 to 5 show examples in which we consider isotropic beams with $\mu_x=0.49$ and $M=0.1$. The beam density is determined to give $k_H/k_L=6/5$. With these parameters, $k_H=1.6$, $k_L=1.1$, $\delta a_H/\delta b_H=1$, $\chi=1$, $K=1.55$ and $\mu_y=0.49$. In Fig. 3, the breathing mode oscillation is selectively excited ($\Gamma=1$). In this case, large 2:1 resonance islands are present. The particles locked into the islands gain excess energy through resonant interaction with the core and become halos. In Fig. 4, the quadrupole mode oscillation is selectively excited ($\Gamma=-1$). Though 2:1 resonance islands are also observed in this case, the resonance width is smaller than that in the pure breathing mode case. Figure 5 shows the results for the mixed case where two normal modes are equally excited ($\Gamma=0$). In the mixed case, the resonance overlap of two 2:1 particle-core resonances occurs, and it results in global chaos which clearly seen in Fig. 3(b). The strong chaosity is expected to enhance the chance for the particles initially located just outside the core to be trapped by the 2:1

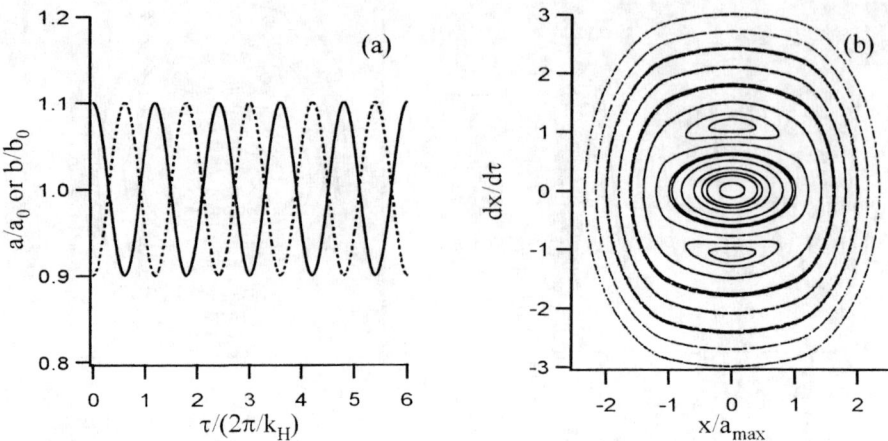

FIGURE 4. Results for an isotropic beam (pure quadrupole mode case). The parameters are set to be $\zeta=1$, $\eta=1$, $\mu_x=0.49$, $M=0.1$, and $\Gamma=-1$. (a) Time-evolution of the beam core. The solid and broken lines respectively represents the horizontal and vertical core half-width scaled by the matched one. (b) Poincaré plots for test particles. The abscissa is the horizontal coordinate scaled by the maximum horizontal core width a_{max}.

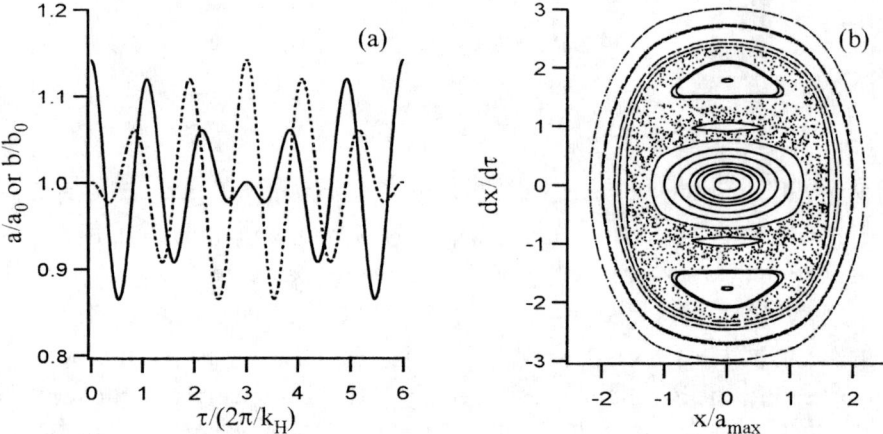

FIGURE 5. Results for an isotropic beam (mixed case). The parameters are set to be $\zeta=1$, $\eta=1$, $\mu_x=0.49$, $M=0.1$, and $\Gamma=0$. (a) Time-evolution of the beam core. The solid and broken lines respectively represents the horizontal and vertical core half-width scaled by the matched one. (b) Poincaré plots for test particles. The abscissa is the horizontal coordinate scaled by the maximum horizontal core width a_{max}.

resonance and become halos. It is worth mentioning that the strong chaos is induced in this case with relatively small mismatch of 10%.

Anisotropic Cases

Second, we will consider anisotropic cases. As an example, we show, in Figs. 6 to 8, the time-evolution of core oscillation and the Poincaré surface of section plots of test particles for an anisotropic beam with $\mu_x=0.50$, $\zeta=1.42$, $\eta=1$ and $M=0.1$. The parameters are determined to give $k_H/k_L=8/5$. With these parameters, $k_H=2.33$, $k_L=1.46$, $\delta a_H/\delta b_H=0.13$, $\chi=0.71$, $K=1.28$ and $\mu_y=0.69$. The most interesting feature seen in these figures is the regularity of the test particle motion in the mixed case (see Fig. 8), which is striking contrast to the isotropic cases (7). The reason why no strong chaosity is observed in the mixed case is that some of two 2:1 particle-core resonances do not occur with these parameters. As readily seen in Figs. 5 and 6, in the vertical plane, the 2:1 resonance islands are observed in the pure breathing mode case ($\Gamma=1$), but not in the pure quadrupole mode case ($\Gamma=-1$). On the contrary, in the horizontal plane, the 2:1 resonance islands are present in the pure quadrupole mode case ($\Gamma=-1$), but not in the pure breathing mode case ($\Gamma=1$). The absence of some of 2:1 resonances leads to the suppression of the resonance overlap of two 2:1 resonances, which results in the suppression of strong chaosity in the mixed case. The regularity of the test particle motion strongly suggests that halo intensity in anisotropic beams can be reduced by avoiding the resonance overlap of two 2:1 resonances.

DISCUSSION

The next question is how we can choose appropriate parameters to suppress the resonance overlap. To find the criterion to choose parameters, we need to know the necessary conditions for the existence of 2:1 particle-core resonance. The necessary conditions for the existence of 2:1 resonance in the horizontal plane are

$$2\mu_x < k_H < 2 \quad \text{(for high-frequency mode)}, \tag{21}$$

and

$$2\mu_x < k_L < 2 \quad \text{(for low-frequency mode)}. \tag{22}$$

Analogously, those in the vertical plane are

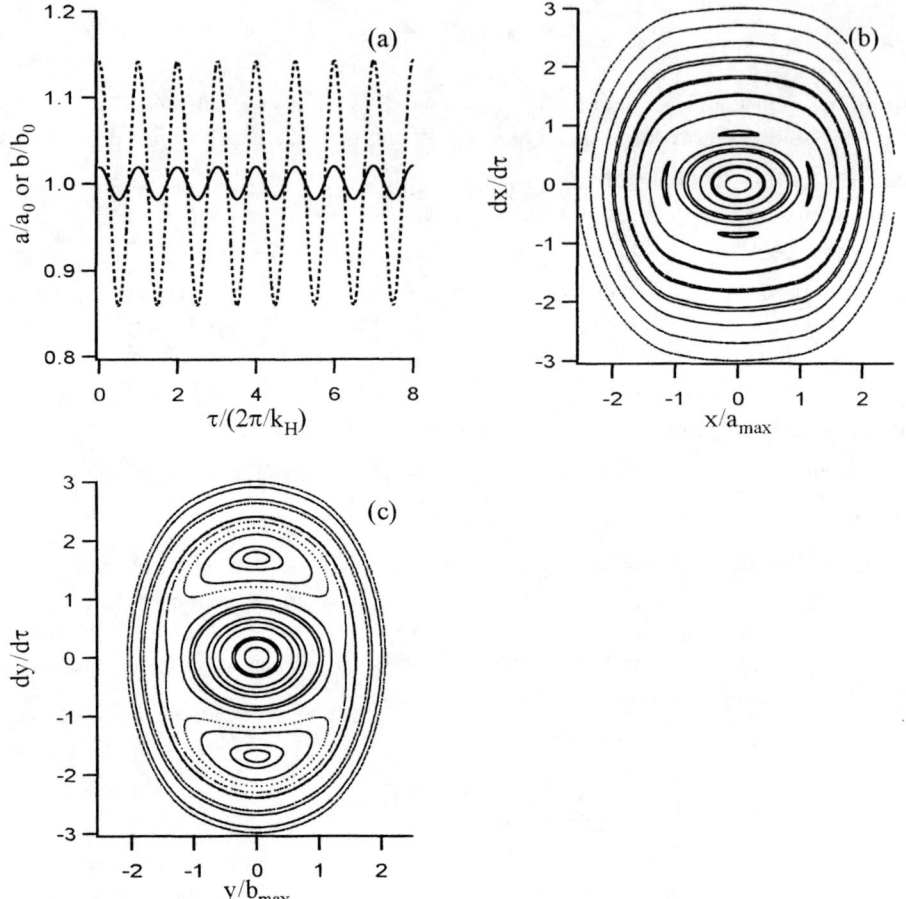

FIGURE 6. Results for an anisotropic beam (pure breathing mode case). The parameters are set to be $\zeta=1.42$, $\eta=1$, $\mu_x=0.50$, $M=0.1$, and $\Gamma=1$. (a) Time-evolution of the beam core. The solid and broken lines respectively represents the horizontal and vertical core half-width scaled by the matched one. (b) Poincaré plots for test particles in the horizontal direction. The abscissa is the horizontal coordinate scaled by the maximum horizontal core width a_{max}. (c) Poincaré plots for test particles in the vertical direction. The abscissa is the vertical coordinate scaled by the maximum vertical core width b_{max}.

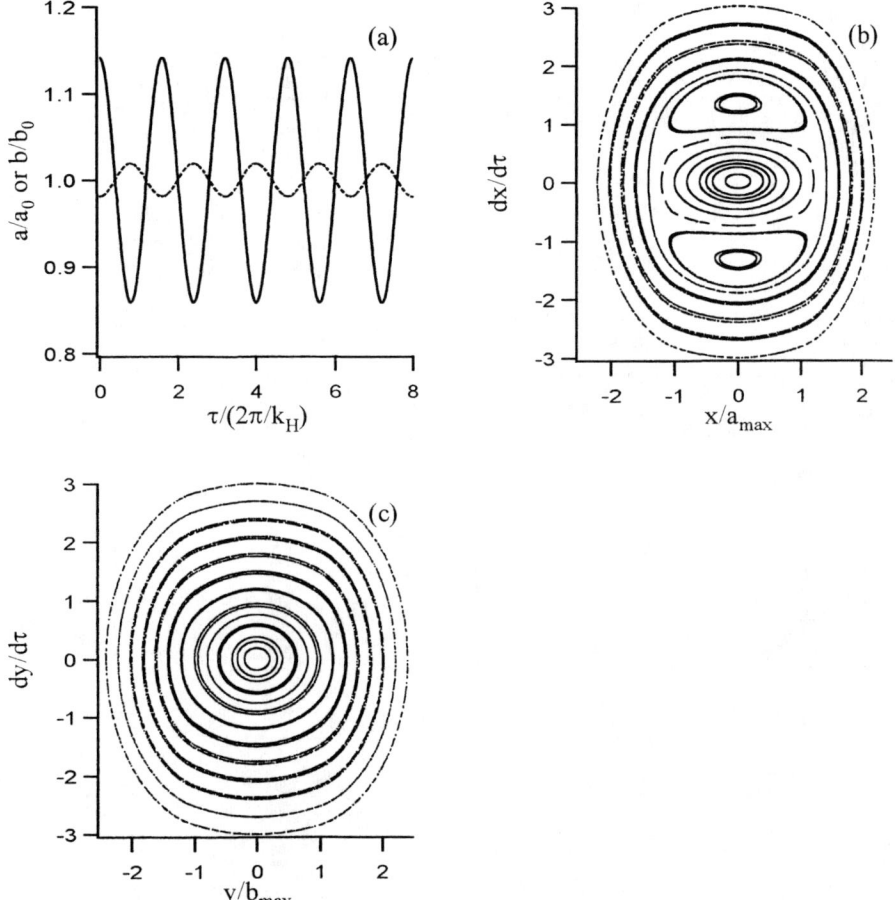

FIGURE 7. Results for an anisotropic beam (pure quadrupole mode case). The parameters are set to be $\zeta=1.42$, $\eta=1$, $\mu_x=0.50$, $M=0.1$, and $\Gamma=-1$. (a) Time-evolution of the beam core. The solid and broken lines respectively represents the horizontal and vertical core half-width scaled by the matched one. (b) Poincaré plots for test particles in the horizontal direction. The abscissa is the horizontal coordinate scaled by the maximum horizontal core width a_{max}. (c) Poincaré plots for test particles in the vertical direction. The abscissa is the vertical coordinate scaled by the maximum vertical core width b_{max}.

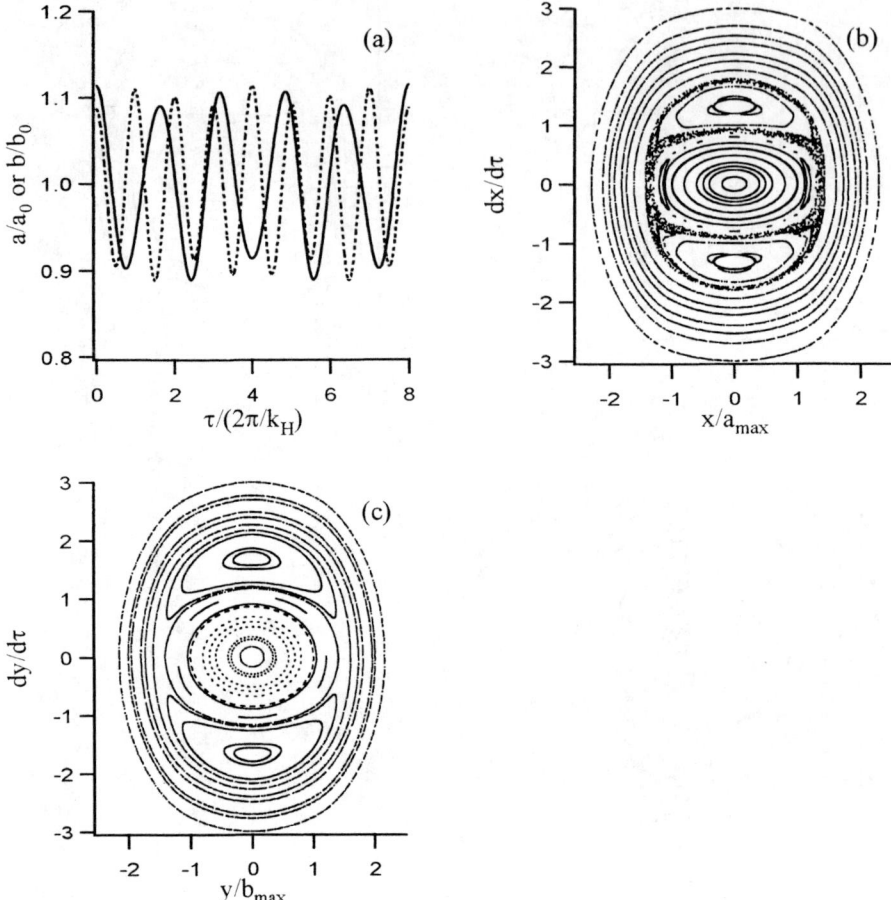

FIGURE 8. Results for an anisotropic beam (mixed case). The parameters are set to be $\zeta=1.42$, $\eta=1$, $\mu_x=0.50$, $M=0.1$, and $\Gamma=0$. (a) Time-evolution of the beam core. The solid and broken lines respectively represents the horizontal and vertical core half-width scaled by the matched one. (b) Poincaré plots for test particles in the horizontal direction. The abscissa is the horizontal coordinate scaled by the maximum horizontal core width a_{max}. (c) Poincaré plots for test particles in the vertical direction. The abscissa is the vertical coordinate scaled by the maximum vertical core width b_{max}.

$$2\mu_y < \frac{k_H}{\zeta} < 2 \qquad \text{(for high-frequency mode)}, \qquad (23)$$

and

$$2\mu_y < \frac{k_L}{\zeta} < 2 \qquad \text{(for low-frequency mode)}, \qquad (24)$$

where μ_y is the tune depression for the vertical motion. It should be noted that all these conditions are satisfied in isotropic cases with any beam density as readily seen in Fig. 1. It means that the resonance overlap of two 2:1 particle-core resonances occurs in isotropic situations if two normal modes are simultaneously excited. On the contrary, some of these conditions are not satisfied generally in anisotropic situations. For example, in the case of Figs. 6 to 8, Eqs. (22) and (23) are satisfied while Eqs. (21) and (24) are not satisfied. This gives us a practical criterion to choose parameters in designing high-intensity synchrotrons and storage rings.

SUMMARY

The effects of simultaneous excitation of two normal modes on beam halo formation have been studied based on the particle-core model. In the analysis, we have considered the isotropic and anisotropic coasting beams. In isotropic cases, two 2:1 particle-core resonances exist, and the overlap of these two resonances causes strong chaos in cases where two normal modes are simultaneously excited. The strong chaos which is expected to increase halo intensity is induced in wider parameter space than in the cases where one of two normal modes is selectively excited. It suggests that the condition for precise matching in an actual machine to suppress halos is severer than that estimated from the pure breathing or pure quadrupole mode oscillation cases. On the contrary, it has been found that the overlap of these two resonances can be avoided by introducing appropriate anisotropy. The chaotic motion of test particles can be suppressed by avoiding the resonance overlap, which strongly suggests that halo intensity can be reduced by choosing appropriate parameters in anisotropic situations. These results give us a practical criterion in designing a high-intensity ion synchrotron and storage ring.

ACKNOWLEDGMENTS

The author gratefully acknowledges helpful and fruitful discussions with Dr. S. Machida of KEK on several points in this paper.

REFERENCES

1. O'Connell, J. S., Wangler, T. P., Mills, R. S., and Crandall, K. R., "Beam Halo Formation from Space-Charge Dominated Beams in Uniform Focusing Channel", in *Proceedings of the 1993 Particle Accelerator Conference*, 1993, p. 3926; Lagniel, J. M., Nucl. Instrum. Methods A **345**, 46 (1994); Gluckstern, R. L., Phys. Rev. Lett. **73**, 1247 (1994); Riabko, A., Ellison, M., Kang, X., Lee, S. Y., Li, D., Liu, J. Y., Pei, X., and Wang, L., Phys. Rev. E **51**, 3529 (1995); Okamoto, H., and Ikegami, M., Phys. Rev. E **55**, 4694 (1997); Wangler, T. P., Crandall K. R., Ryne, R., and Wang, T. S., Phys. Rev. ST Accel. Beams **1**, 084201 (1998); Ikegami, M., "Particle-Core Analysis of Mismatched Beams in a Periodic Focusing Channel", to be published in Phys. Rev. E.
2. Lagniel, J. M., Nucl. Instrum. Methods A **345**, 405 (1994); Qian, Q., Davidson, R. C., and Chen, C., Phys. Rev. E **51**, R5216 (1995); Fink, Y., Chen, C., and Marable, W. P., Phys. Rev. E **55**, 7557 (1997).
3. Kapchinskij, I. M., and Vladimirskij, V. V., "Limitations of Proton Beam Current in a Strong Focusing Linear Accelerator Associated with the Beam Space Charge", in *Proceedings of International Conference for High Energy Accelerators*, 1959, p. 274.
4. Reiser, M., *Theory and Design of Charged Particle Beams*, John Wiley & Sons, 1994, Chap. IV.
5. For example, Kellog, O. D., *Foundation of Potential Theory*, Frederick Ungar Publishing Company, 1953, Chap. VII.
6. Forest, E., and Ruth, R. D., Physica D **43**, 105 (1990); Yoshida, H., Phys. Lett. A **150**, 262 (1990).
7. Although we can see, in Fig. 8(b), weak chaosity of test particle motion which is induced by the overlap of 2:1 and 4:1 particle-core resonances, the overlap can be avoided adopting a slightly different set of parameters. In avoiding the resonance overlap, the chaosity of test particle motion can be suppressed to a level comparable to Fig. 8(c).

Numerical Simulation of Multicomponent Ion Beams in Transport Lines

V. Alexandrov[*], Yu. Batygin[#], N. Kazarinov[*], V. Shevtsov[*], G. Shirkov[*]

[*]*Joint Institute for Nuclear Research (JINR), 141980, Dubna, Russia*
[#]*Institute for Physical and Chemical Research (RIKEN), Saitama, 351-01, Japan*

Abstract. A program library for numerical simulation of a multicomponent charged particle beam in transport lines is presented. The library is aimed for simulation of high current, low energy multicomponent ion beam from ion source through beamline and realized under the Windows user interface for the IBM PC. It is used for simulation and optimization of beam dynamics and based on successive and consistent application of two methods: the momentum method of distribution function (RMS technique) and particle in cell method.

The library has been used to simulate and optimize the transportation of tantalum ion beam from the laser ion source through the Low Energy Beam Transport line (LEBT) into the ionic RFQ linac at CERN.

INTRODUCTION

The program library for numerical simulation of the charged particle beam transportation lines realized on the IBM PC with Windows user interface of Visual Basic is presented. A number of computer codes for the numerical simulation of charged particle beams with non-linear space charge field of the beam have been developed since 70-s[1-8].

The present program library is used for the numerical calculation and optimization of beam dynamics in the transport lines including various magnetic and electrostatic elements. Nonlinear transverse space charge is taken into account. The main advantage of the code is simultaneous simulation of a number of particle components with different masses and charge states. This peculiarity is very important, in particular, for the low energy stage transportation of a high current ion beam from ion source to accelerator or beam analyzing system.

The library is based on three different methods of mathematical simulation: (i) the momentum method for particle distribution function[9,10], (ii) macroparticle method or particle-particle method (PP-method) and (iii) the solution of Poisson equation in 2D Cartesian Coordinates using the fast Fourier transformation method[11].

Fast analysis and study of the averaged beam characteristics, such as mean velocity and root-mean-square (RMS) dimensions is attained by the momentum method. The main advantages of the momentum method for distribution function are simplicity and applicability for beam transport optimization in comparison with methods of multiparticle description.

The PP-Method is successfully used for detailed investigation of beam distribution function. This method is also helpful to consider the nonlinearity of electron and ion space charge fields.

Earlier these methods were applied to calculate and optimize the high intensity beam injection line for the Meson Factory of the Institute for Nuclear Research (Troitsk)[12]. Those methods and codes were used as a background for the new generation code library. The first version of the present code library without using the fast Fourier transformation method for the solution of Poisson equation was presented earlier[13,14].

The library is installed on the modern PC hardware and software to extend a number of used finite particles and sells of field mesh in simulation as well as available transportation line elements. An advanced Windows graphical interface makes it comfortable and friendly for the User in the interactive mode operation. The program may be used for numerical simulation of dynamics of multicomponent beams with a realistic charged state distribution.

MOMENT METHOD USE FOR BEAM TRANSPORT STUDY

The distribution of particles within the phase space of coordinate $\vec{x}$ and velocities $\vec{v}$ is described by the distribution function $f(\vec{x}, \vec{v}, t)$. In consideration of the particles' electromagnetic fields and of impact processes which take place in ion sources and beams, the movement of charged particles is described in a generalized form by a system of self-consistent kinetic equations completed by Maxwell equations. Solution of the complete set of these equations meets great difficulties. In practice several approaches and numerical models are applied to solve these equations.

The method of complete moments of the distribution function is one of the most successful approaches to solving the Vlasov equations. The fundamentals of this method are established in the Ref.[15] and the corresponded problems related to multicomponent electron-ion beams were considered in the Ref.[16]. If the distribution function is integrated in the phase space of coordinates $\vec{x}$ and velocities $\vec{v}$ with an arbitrary power of coordinates and velocities then the moments of the distribution function can be introduced. The moment of zero order

$$N = \int f \, d\vec{x} \, d\vec{v}$$

equals to the total number of particles in the beam. The moments of the first order

$$\bar{x}_i = \int f x_i \, d\vec{x} \, d\vec{v}, \quad \bar{v}_i = \int f v_i \, d\vec{x} \, d\vec{v}$$

describe the location of the mass centre and mean velocities.

$$\bar{x}_i^2 = \int f x_i^2 \, d\vec{x} \, d\vec{v}, \quad \bar{v}_i^2 = \int f v_i^2 \, d\vec{x} \, d\vec{v}$$

describe the RMS beam size and velocities.

Let us define a second order moments matrix M of the distribution function according to Ref.[9] while considering the field nonlinearity:

$$M^{II} = \begin{pmatrix} M_{xx} & M_{xv} \\ M_{xv}^* & M_{vv} \end{pmatrix} \quad (1)$$

$$M_{xx}^{i,j} = \int f x_i x_j d\vec{x} d\vec{v}, \qquad M_{xv}^{i,j} = \int f x_i v_j d\vec{x} d\vec{v}$$

$$M_{vv}^{i,j} = \int f v_i v_j d\vec{x} d\vec{v}, \quad i, j = 1,2 \quad (2)$$

Here f is the beam distribution function, and x_i, v_j are particle coordinates and velocities. The evolution of the matrix M under beam transportation through linear electromagnetic fields is defined by the set of equations is[10]:

$$\frac{dM_{xx}}{dz} = M_{xv} + M_{xv}^*,$$

$$\frac{dM_{xv}}{dz} = M_{vv} + M_{xx} B^* + M_{xv}^* A^*, \quad (3)$$

$$\frac{dM_{vv}}{dz} = BM_{xv} + M_{xv}^* B^* + AM_{vv} + M_{vv} A^*$$

where $z = v_z t$ and $v_z = \beta c$ are longitudinal coordinates and velocity. Matrix B includes external and space charge forces, matrix A includes external forces only (magnetic field defined by particle transverse motion is neglected), matrixes M^*, A^*, B^* are transposed matrices M, A, B. The matrices A and B for typical magnetic and accelerating cavities have the following forms:

- *Solenoid with axial magnetic field B:*

$$B^{ext} = \frac{Ze}{2\beta\gamma Am_p c^2} \frac{\partial B_x(z)}{\partial z} \begin{pmatrix} 0 & 1 \\ -1 & 0 \end{pmatrix}, \quad A^{ext} = \frac{ZeB_z(z)}{\beta\gamma Am_p c^2} \begin{pmatrix} 0 & 1 \\ -1 & 0 \end{pmatrix} \quad (4)$$

- *Quadrupole with gradient G:*

$$B^{ext} = \frac{ZeG}{\beta\gamma Am_p c^2} \begin{pmatrix} -1 & 0 \\ 0 & 1 \end{pmatrix}, \qquad A^{ext} = 0, \quad (5)$$

- *Horizontal bending magnet (sector type):*

$$B^{ext} = -\left(\frac{ZeB_z}{\beta\gamma Am_p c^2}\right)^2 \begin{pmatrix} 1 & 0 \\ 0 & 0 \end{pmatrix}, \quad A^{ext} = 0, \tag{6}$$

- *Vertical bending magnet (sector type):*

$$B^{ext} = -\left(\frac{ZeB_z}{\beta\gamma Am_p c^2}\right)^2 \begin{pmatrix} 0 & 0 \\ 0 & 1 \end{pmatrix}, \quad A^{ext} = 0, \tag{7}$$

- *Accelerating cavity:*

$$A^{ext} = \frac{1}{\gamma}\frac{\partial\gamma}{\partial z}\begin{pmatrix} 1 & 0 \\ 0 & 1 \end{pmatrix}. \tag{8}$$

The elements of moment matrix M for the horizontal bending magnet with the pole face rotation α and bending radius ρ are transformed accordingly:

$$M = W^* M W,$$

$$W = \begin{pmatrix} 1 & 0 & 0 & 0 \\ 0 & 1 & 0 & 0 \\ \frac{\tan\alpha}{\rho} & 0 & 1 & 0 \\ 0 & -\frac{\tan\alpha}{\rho} & 0 & 1 \end{pmatrix} \tag{9}$$

In matrix elements for a vertical bending magnet the value of α is changed for $-\alpha$. In formulas (4)-(8) the values Ze and Am_p are ion charge and mass respectively (for electrons $Z=1$ and $Am \to m_e$, where m_e is the electron mass), γ is the relativistic factor.

Space charge forces in (3) are considered to be linear[10,15]. In this case:

$$B^s = \frac{I}{\beta^2\gamma^3 I_A}\frac{M_{xx}^{-1/2}}{Sp M_{xx}^{1/2}} \tag{10}$$

wher $I_A = \beta\gamma Am_p / Ze^2$ is the Alfven current. Matrix $M_{xx}^{1/2}$ is defined with conditions $M_{xx}^{1/2} M_{xx}^{1/2} = M_{xx}$, $M_{xx}^{1/2} M_{xx}^{-1/2} = E$, where E is the unit matrix. The following expressions connect the elements of the matrices $M^{1/2}$ and $M^{-1/2}$:

$$M_{11}^{1/2} = \frac{M_{11} + \left[\det M_{xx}\right]^{1/2}}{\left(Sp M_{xx} + 2(\det M_{xx})^{1/2}\right)^{1/2}}, \quad M_{22}^{1/2} = \frac{M_{22} + \left[\det M_{xx}\right]^{1/2}}{\left(Sp M_{xx} + 2(\det M_{xx})^{1/2}\right)^{1/2}},$$

$$M_{12}^{1/2} = \frac{M_{12} + \left[\det M_{xx}\right]^{1/2}}{\left(Sp M_{xx} + 2(\det M_{xx})^{1/2}\right)^{1/2}}, \quad M_{12}^{1/2} = M_{21}^{1/2}, \qquad (11)$$

$$M_{11}^{-1/2} = \frac{M_{22}^{1/2}}{\det(M^{1/2})}, \quad M_{12}^{-1/2} = -\frac{M_{12}^{1/2}}{\det(M^{1/2})}, \quad M_{22}^{-1/2} = \frac{M_{11}^{1/2}}{\det(M^{1/2})}, \quad M_{21}^{-1/2} = M_{12}^{-1/2}$$

The solution of set of equations (3) is used to define the rms values of the beam (transverse beam sizes and velocities, cross term). The main advantage of this approach is a short calculation time for the main beam characteristics.

Approximation of self linear forces is correct for long beams with the constant transverse density of charged particles. Real particle distributions usually have nonlinear space charge fields that could be described with higher order moments and, according to this, a higher order set of equations. The approach of effective linearisation of transversal fields for a special kind of distribution function was introduced to solve the problem in the frames of second order moments[10].

EQUATIONS OF PARTICLE MOTION IN PP-METHOD

The macroparticle method allows us to study the detailed beam space characteristics, the distribution function of particles, including the nonlinear self and external fields. This method is the most general and powerful one for the simulation of the motion of continuous medium, gasdynamics, dynamics of charged particle beams.

At the initial moment of time, the phase space volume of each medium component is divided into a certain number of non superimposing elementary volumes or cells and the motion of such volume is identified with the motion of certain finite particle with the respective charge and mass. These model particles are finite particles or macroparticles. In order to get an adequate information on the considered process, it is necessary to examine a sufficiently great number of particles and to statistically average results of simulation. Initial positions of particles or coordinates, i.e. starting points of the phase trajectories are selected by statistical methods according to the given initial distribution function. Numbers of finite particles and cells of mesh determine the capacity, accuracy and resolution of simulation procedure. These values are limited usually by the power of computer.

The set of equations of the particle motion in various external fields and self space charge field has the following form:

$$\frac{d^2 x_i}{ds^2} = F_{xi}^p + \frac{Q}{\gamma^2 n} \sum_{j=1}^{n} \frac{x_i - x_j}{r_{ij}^2} Z_j$$

$$\frac{d^2 y_i}{ds^2} = F_{yi}^p + \frac{Q}{\gamma^2 n} \sum_{j=1}^{n} \frac{y_i - y_j}{r_{ij}^2} Z_j,$$

(12)

where x_i, y_i are transverse coordinates for i's particle, $Q = 2I/\beta^3 \gamma I_\alpha$ is the space charge parameter, I is the linear current density of the beam, $\beta = v_z/c$ is the relative longitudinal velocity of any particle, c is the velocity of light, $I_\alpha = m_p c^3/e$ for ions and $I_\alpha = m_e c^3/e$ for electrons. The distance r_{ij} between particles i and j is: $r_{ij}^2 = (x_i - x_j)^2 + (y_i - y_j)^2$ if $r_{ij} \geq 2 a_0$ and $r_{ij} = 2 a_0$ if $r_{ij} < 2 a_0$, where a_0 is the radius of the particle and n is the particle number. The forces F_{xi}^p and F_{yi}^p determine the particle interaction with external electromagnetic fields:

- *for an axial symmetric magnetic field of any solenoid with induction* $B_z(z)$, *then*

$$F_{xi}^s = Q_1 \left(\frac{\partial y_i}{\partial z} B_z + \frac{1}{2} \frac{\partial B_z}{\partial z} y_i \right),$$

$$F_{yi}^s = -Q_1 \left(\frac{\partial x_i}{\partial z} B_z + \frac{1}{2} \frac{\partial B_z}{\partial z} x_i \right),$$

(13)

- *for a quadrupole with the gradient of the magnetic field G(z):*

$$F_{xi}^Q = -Q_1 G(z) x_i,$$

$$F_{yi}^Q = -Q_1 G(z) y_i,$$

(14)

- *for a vertical bending magnet with the amplitude* B_y *(sector type):*

$$F_{xi}^M = -Q_1^2 B_y^2 x_i,$$

$$F_{yi}^M = 0,$$

(15)

- *for a horizontal bending magnet with the amplitude* B_x *(sector type):*

$$F_{xi}^M = 0,$$

$$F_{yi}^M = -Q_1^2 B_x^2 y_i.$$

(16)

Here $Q_1 = Ze/Am_p c^2 \beta$ for ions and $Q_1 = e/m_e c^2 \beta$ for electrons. In these equations Ze is the total ion charge, m_p, m_e are the masses of proton and electron, respectively, A is the atomic number of the ion. The values of the forces are zero $F_{xi}^{fs} = F_{yi}^{fs} = 0$ in free space.

LINEAR TRANSFORMATION OF THE PHASE SPACE COORDINATES

Because of finite number of the macro particles N the r.m.s. dimensions of the beam in the phase space are not exactly equal to the modeling numbers (relative error is about $N^{-1/2}$). Besides these variables are changed for other random generation of the macro particles. For achieving the coincidence of the initial conditions in the averaged sense for the various random generation the linear transformation of the phase space variables (x, x', y, y') may be applied:

$$\begin{pmatrix} x_f \\ x'_f \end{pmatrix} = R \begin{pmatrix} x_i \\ x'_i \end{pmatrix},$$

where R is 2 x 2 matrix, index "i" denotes the value of the parameter after random generation. And the same for y.

Let us introduce averaged emittance and Twiiss's parameters of the beam:

$$\varepsilon^2 = \overline{x^2}\,\overline{x'^2} - \left(\overline{xx'}\right)^2,\ \beta = \overline{x^2}/\varepsilon,\ \alpha = -\overline{xx'}/\varepsilon,\ \gamma = \overline{x'^2}/\varepsilon.$$

The upper bar means the averaging with macro particles. With these definitions the matrix R has the form:

$$R = \sqrt{\frac{\varepsilon_f}{\varepsilon_i}} \begin{pmatrix} \sqrt{\frac{\beta_f}{\beta_i}} & 0 \\ \frac{(\alpha_f - \alpha_i)}{\sqrt{\beta_f \beta_i}} & \sqrt{\frac{\beta_i}{\beta_f}} \end{pmatrix}$$

Index "f" denotes the desirable value of the parameter. This matrix corresponds to the transfer matrix of the accelerating / decelerating structure with the phase advance of the betatron oscillation to be equal to 2π. The (y, y') phase coordinates is transformed by the same rule.

CALCULATION OF SPACE CHARGE FORCES IN 2D CARTESIAN COORDINATES

Space charge field of z-uniform beam is calculated from the Poisson equation in two-dimensional Cartesian coordinates:

$$\frac{\partial^2 U}{\partial x^2} + \frac{\partial^2 U}{\partial y^2} = -Q(x,y), U(\Gamma) = 0,$$

with Dirichlet boundary condition for potential U at the surface of an infinite pipe, Γ.

Unknown potential of the beam at grid points, $U(i, j)$, is represented as Fourier series:

$$U_{ij} = \sum_{n=1}^{N-1}\sum_{m=1}^{N-1} \overline{U}_{nm} \sin\left(\frac{\pi n i}{N_x}\right) \sin\left(\frac{\pi m i}{N_y}\right),$$

similar for space charge density, $Q(x, y)$. After substitution of these expansions into Poisson's equation, the Fourier coefficients of space charge and potential are connected by algebraic relationship:

$$\overline{U}_{nm} = \frac{Q_{nm}}{\left(\frac{\pi n}{a}\right)^2 + \left(\frac{\pi m}{b}\right)^2},$$

giving solution to the problem. Here a and b are sizes of the chamber in horizontal and vertical directions, respectively. The inverse Fourier transformation gives us the values of beam potential in the knots of the grid. The values of Coulomb forces acting in horizontal and vertical directions are calculated as central differences of potential:

$$F_x = \frac{U(i+1, j) - U(i-1, j)}{\left(a/(N_x + 1)\right)}, F_y = \frac{U(i, j+1) - U(i, j-1)}{\left(b/(N_y + 1)\right)}.$$

BEAM INITIAL DATA FOR THE DISTRIBUTION FUNCTION IN MOMENTS AND PP-METHODS

Initial positions of particles in phase space are determined by the values of Twiss parameters $\alpha_x, \alpha_y, \beta_x, \beta_y$, by the values of emittances of the beam $\varepsilon_x, \varepsilon_y$ in transverse phase planes xv_x, yv_y and by type of distribution function. To unify the initial data, the RMS-dimensions and RMS-velocities of the beam are defined as follows:

$$\overline{a_x^2} = \varepsilon_x \beta_x, \overline{a_y^2} = \varepsilon_y \beta_y, \overline{xv_x} = -\varepsilon_x \alpha_x, \overline{yv_y} = -\varepsilon_y \alpha_y,$$

$$\overline{v_x^2} = \left(\overline{xv_x}\right)^2 + \left(\overline{yv_y}\right)^2) / \overline{a_x^2}, \overline{v_y^2} = \left(\overline{xv_x}\right)^2 + \left(\overline{yv_y}\right)^2) / \overline{a_y^2}$$

The beam parameters are:
- total current (µA);
- type of beam: electron or ion; the beam in the first case has a charge, which is equal to the electron charge, in the other case, the beam is characterized by the ratio of the ion charge to the atomic number; and it is necessary to set the atomic number here.
- type of density distribution in the four-dimensional phase space: Kapchinsky-Vladimirsky distribution is used for the method of moments.

Fig. 1 shows an example of User interface to determinate the beam parameters. This form appears in the monitor screen when User chooses the item "Initial Data" + "Beam's Parameters+Initial Data" in the main menu.

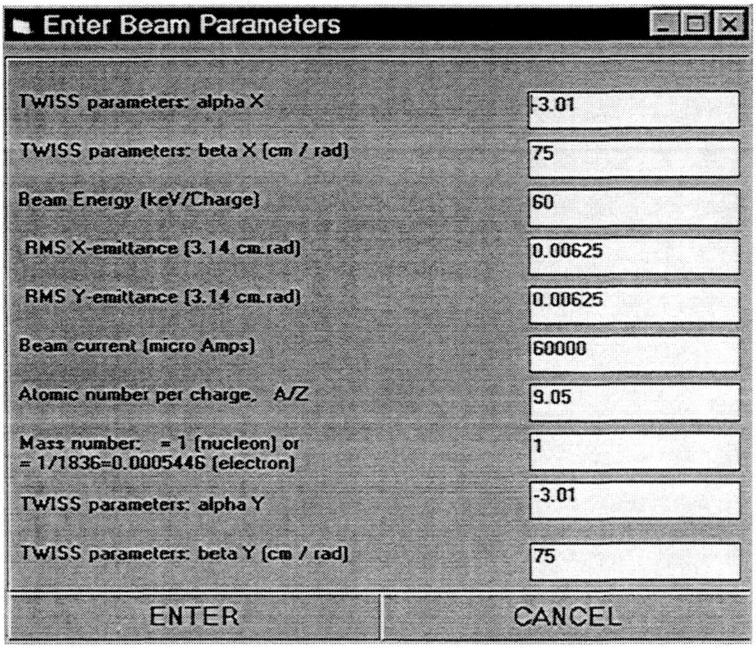

FIGURE1. The form for input of beam parameters

The initial data in PP-method are determined by the same values. There are four types of density distribution in the four-dimensional phase space: uniform, gaussian, water bag and parabolic distributions. It is possible to select one of them according to the best approximation of the conditions to solve this problem. The initial distribution of the particles in four-dimensional phase space, i. e. it's coordinates and velocities are calculated by special programmes using the random number generators[17]. All of the above beam parameters are determined when the item "Initial Data" + "Beam's Parameters + Initial Data" is chosen in the main menu. A special form appears in the screen of the monitor as a result of choosing this item. One can see this form in Fig. 2.

User can determine the beam data by focusing the screen pointer to the text field and putting the needed values from the keyboard. The shift to the next text field is produced by clicking the key "Tab". It is necessary to click button "Cancel" and get rid of these data. The beam data will be loaded into the memory when User selects the button "Enter". This data may be written in a special file (see below).

The PP-method makes it possible to take into account a charge distribution and Coulomb interaction of all species. In this case the total current is distributed between different species if the beam has more than one ion species. The charge distribution is determined in a special file under the name SPREAD.TXT and could be corrected with any text editor. This distribution is plotted in the monitor screen and one can choose a

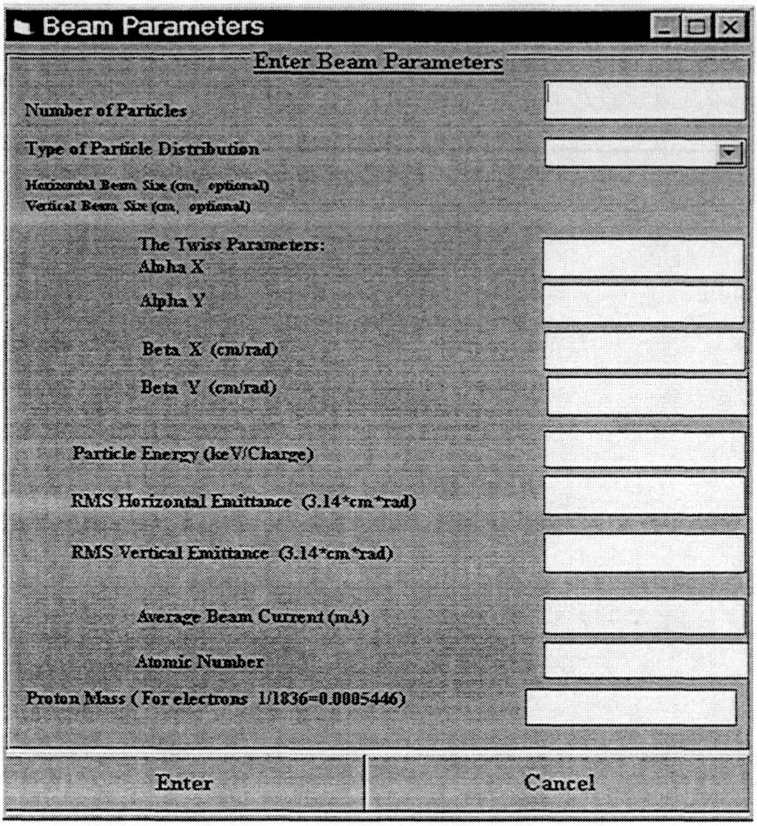

FIGURE 2. The form to select and insert the beam parameters

number of charge states and the lowest and highest states from the complete charge states distribution according to the required accuracy of simulation. Simple mouse clicking is used for all interactive manipulations. The text field with a charge number changes the color when User clicks this field. The next mouse clicking makes the black color over all the region of selected charges. User interface for choosing the charge state is shown in Fig. 3.

This picture could be printed by clicking the button "PRINT DISTRIBUTION" (see Fig. 3). The beam parameters may be written in a special file with extension "*.ind" (initial data) when User chooses the menu item "Initial data" + "Writing of Initial Data". User is able to indicate a path to this file and it's name in the WINDOWS dialogue regime is similar to "File" + "Save As". User can read this file choosing the menu item "Calculation of Channel" + "Reading Initial Data From Disk", if the initial beam parameters are invariable during the next programme runs.

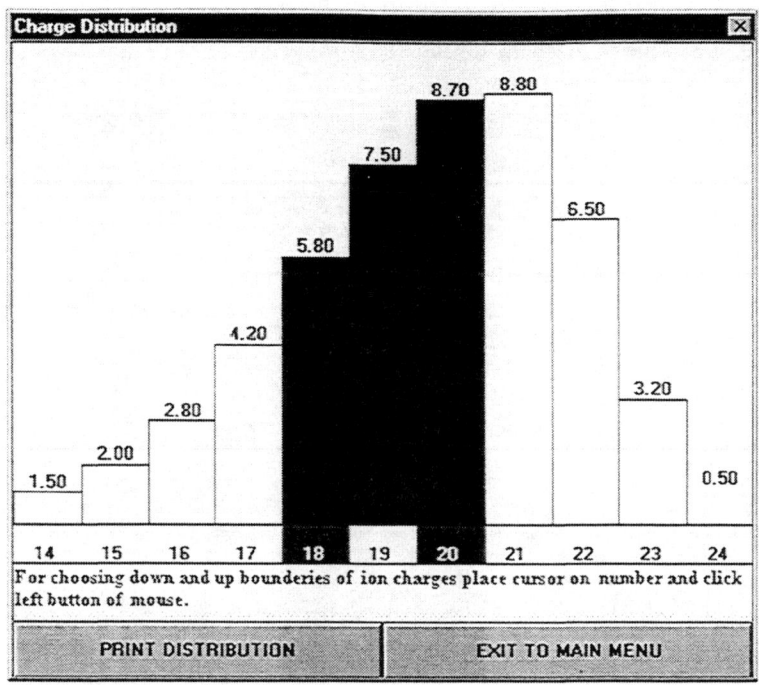

FIGURE 3. User interface for choosing the input charge state distribution

USER INTERFACE FOR SETTING UP PARAMETERS OF THE BEAM TRANSPOR LINE

Parameters of the transportation line are determined in the regime of a dialogue. There are two main forms for this aim.

The first of them allows one to set up the total length of the transportation line, the total number of its elements, the diameter of the inner tube, the z-co-ordinate and radius of the diaphragm. This form is presented in Fig. 4.

User can determine the parameters of the line elements with the second form. First of all, it is necessary to determine the type of elements. Both programmes can work with the following types of elements:
- solenoids,
- vertical dipole magnets,
- horizontal dipole magnets,
- quadrupoles, acceleration cavities,
- free spaces.

When User clicks the button "Enter" (see Fig. 4), then the form which is seen in Fig. 5, appears in the screen of the monitor and offers to User to set on a free space length before this element, the length of the element, it's aperture and value of the physical parameter of this element (the amplitude of the longitudinal field for the solenoid, the value of gradient for the quadrupole and so on).

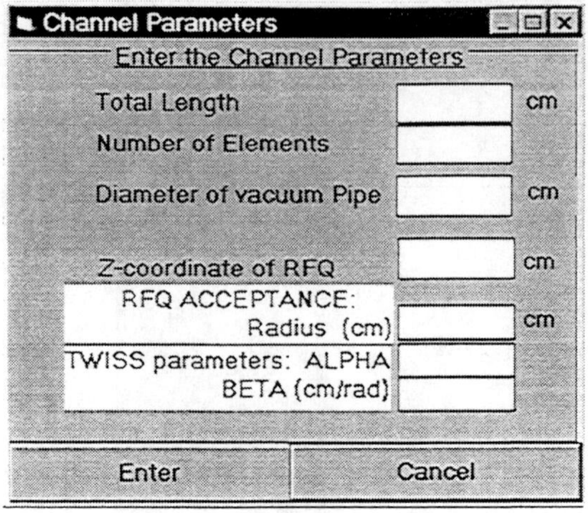

FIGURE 4. User interface for setting up the parameters of transportation line

This dialogue continues until the number of the element is less than the total number of the elements. User can turn down this parameters by clicking the button "Cancel" if parameters of the element are wrong, else he can choose the button "Enter". User can click the button "Information" to get more information about the coincidence of the element parameters. The program automatically returns to the main menu when the parameters of the last element of the transportation line are determined. User can write these data in a special file with extension "*.str" (structure) choosing the item of the main menu "Channel's structure" + "Writing channel's data". The transportation line structure could be restored by choosing the item "Calculation of channel" + "Reading channel's parameters from disk", if it is not necessary to change it during the next runs of the programme.

There are two possibilities to determine the external magnetic field of a solenoid in the programme. The components of the magnetic field are executed analytically when calculation of beam dynamics is fulfilled in paraxial approximation, so it is necessary to set only the amplitude of the longitudinal magnetic field. The components of the field can be set up while reading the file when the components of the magnetic field are represented by the table of values experimentally measured or calculated along the solenoid axis on different radii. Then the dialogue box will appear in the screen of the monitor if User chooses the item of the main menu "Reading data field from file".

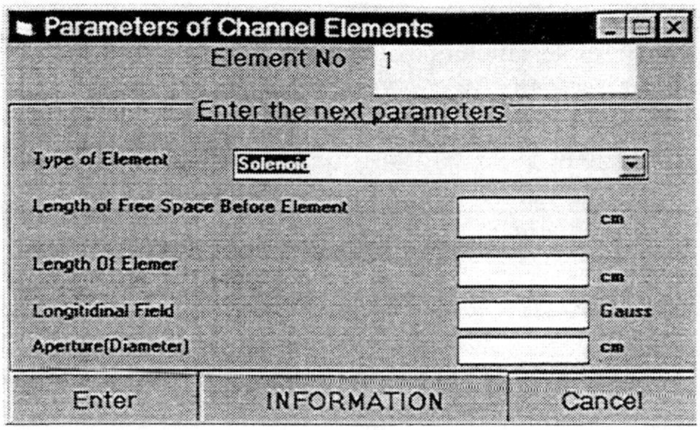

FIGURE 5. User interface to define parameters of the canal element

The file containing the field data with extension "*.dat" (field data) has a structure shown in Table 1:

TABLE 1

R	Z	Br	Bz
r1	z1	Br(r1,z1)	Bz(r1,z1)
r2	z1	Br(r2,z1)	Bz(r2,z1)
....	z1	...	...
r1	z2	Br(r1,z2)	Bz(r1,z2)
r2	z2	Br(r2,z2)	Bz(r2,z2)
....	z2	...	...

The procedure provides linear interpolation of the external magnetic field into the point of a real particle position. The file with the field data is created and corrected in any text editor. All magnetic field values can be changed multiplayed by the scaling factor defined in the start-up form (see Fig. 6).

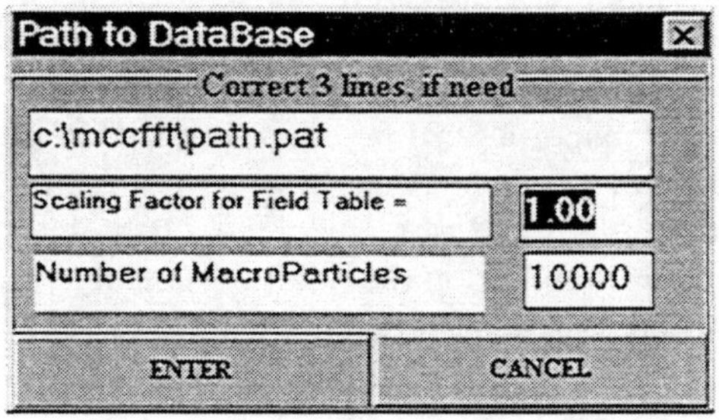

FIGURE 6. The start-up form.

The picture of the channel geometry appears in the screen of the monitor when procedures of setting up parameters of the transportation line and initial beam data are finished. The main initial beam parameters are shown in special windows. Rectangular boxes in the top of this picture represent the elements of the transportation line. The numbers disposed over these boxes show the values of gradients, the amplitude of the magnetic field and so on. An example of this picture is shown below in the Fig. 7 with visualization of beam dynamics in the PP-method.

USER INTERFACE FOR TRANSPORT LINE CALCULATIONS

User interface for the transportation line calculation with the method of moments of the distribution function is described in this paragraph. The transportation line could be calculated in two regimes:
- the transportation beam line is calculated for all it's elements without any stops up to the last element,
- the transportation beam line is calculated element by element with the stop after each element.

User has a possibility to exchange the position of the element along the line and the amplitude of the physical field of this element in the last case. This case will be realised if User chooses the item of the main menu "Channel's Run on Element". The beam envelopes appear on the panel to visualise the beam dynamics calculation in both cases. The red line shows the envelope of the beam in horizontal direction, the blue - in the vertical one. The green line shows the amplitude of the longitudinal magnetic field via the distance.

The calculations go on until the right hand end of the first element is reached. Then the calculations are stopped and the box with the question "Will you continue? Choose the answer!" and two buttons "Yes!" and "No!" appear. The calculations will be continued up to the right hand side of the next element if User chooses "Yes". The background colour of the top panel's part will be changed if User chooses "No". It means that the programme is ready to get new data from User. User can click the mouse pointer on the box of the elements and the foreground colour of this box will be red. If User clicks the button "Insert" on the keyboard, then the box with the number of this element, the old value of the field amplitude in this element will appear in the monitor screen. The empty string offers to User to set up a new value of the amplitude. The calculations will be continued from the right hand end of the previous element or from the beginning of the transportation line when the first element parameters vary then User clicks the mouse pointer in the position of the element.

User is also able to alter the position of any element changing it's field amplitude. It is necessary to click the mouse pointer on the box of this element and click the buttons "→" or "←" for moving the element to the right or to the left. The text box in the bottom of the monitor screen with the longitudinal co-ordinates of the left-hand end of the element will appear in the last case.

OUTPUT DATA

As it was described above, the beam envelopes in horizontal and vertical directions are shown in the monitor screen in the tempo of the transportation line calculation if the method of the moments is selected. The colour traces of each finite particle are presented in the screen and form the image of real beam dimensions along the axes in the PP-method. In this case User can see the output Twiss parameters, the number of lost particles and emittances in the horizontal and vertical directions when the calculations are finished. There are three opportunity for User to have an additional information of the beam parameters:
- the particle losses along the trace of the transportation line,
- the phase portrait and dimensions in the horizontal and vertical dimensions,
- the output current distribution.

User clicks the corresponding button to make a choice (see Fig. 7).

Figures 7-13 represent an example of real beam dynamic simulation during the optimization of Low Energy Beam Transportation (LEBT) line at CERN. The LEBT is aimed to transport the beam from Laser Ion Source (LIS) into the ionic RFQ and consists from two solenoids (two panels on the top of Fig. 7). The data for one of the LIS experimental beams of Tantalum ions with energy of 60 keV per charge were taken to demonstrate here the PP-method possibilities. The total initial ion current of the beam was 60 mA. These calculations used 5000 particles on the 32x32 mesh with the integration step of 0.5 cm. One variant of such kind calculations takes about 1 hour of computation time at Pentium II/300 IBM PC.

The initial charge state distribution of Tantalum ions is shown in Fig. 3 with actual current of each charge state on the top of corresponded ion bar. The input beam parameters are presented in the left column of Fig. 7. This Fig shows the visualization of the beam dynamics. This colour image is a superposition of 5000 traces of all particles. Each charge state is presented with different colour. The right side column shows the output beam parameters. One can see that one third of initial particles (see the lower window in the left column) have been lost in the LEBT for the given beam and cannel parameters.

Figs. 8 and 9 represent the phase portraits in XX' (emittance) and XY (density distribution) for initial and final beam distributions correspondingly. The large ring of particles in the XY plate of Fig. 9 is the image of lost particles that had been excluded from the calculations after reach the radii of inner LEBT tube. The brown ellipse in the left plate and circle in the right plate of Fig. 9 show the RFQ acceptance and aperture.

Different charged states have different conditions of transportation. It is possible to see from the comparison of two final phase plates for Ta^{18+} and Ta^{20+} in Figs. 10 and 11 correspondingly. Particle losses for each charge state along the LEBT are shown in Fig 12. The table in Fig. 13 includes the complete input-output information for every charge states used in consideration. In particular, the penultimate column gives the relative growth of every charge state emittances and the last one indicates the number of particles successfully met the input requirements of RFQ.

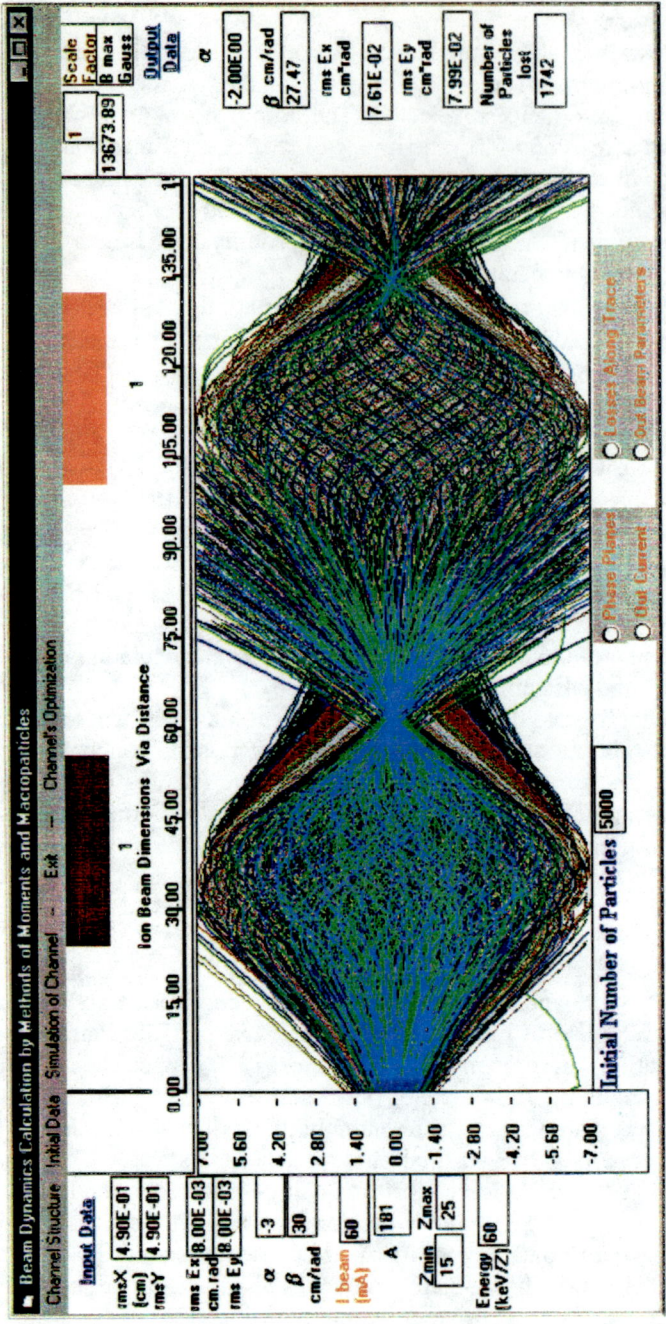

FIGURE 7. Visualization of the beam dynamics simulation

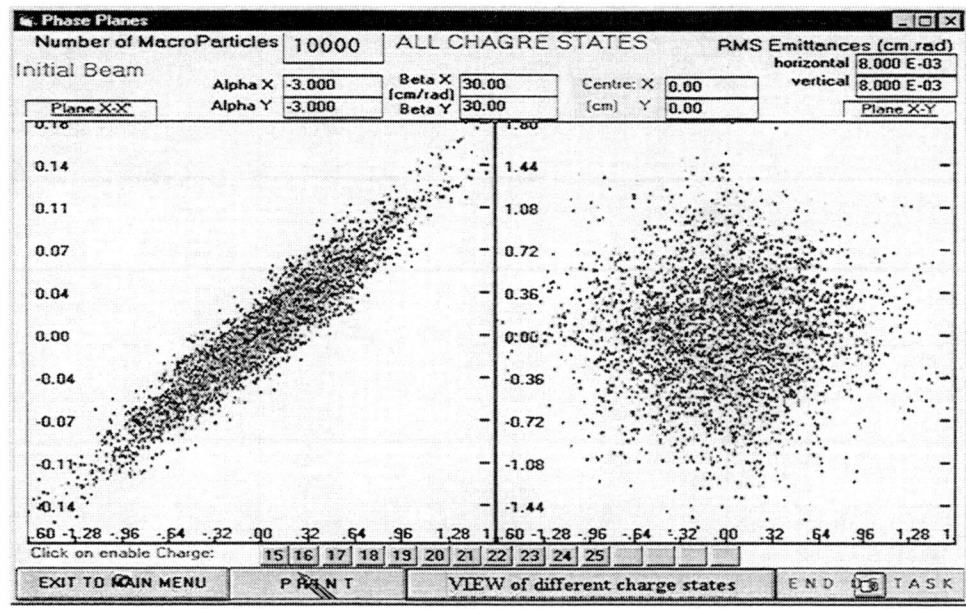

FIGURE 8. Initial phase plates of beam emittance and cross density

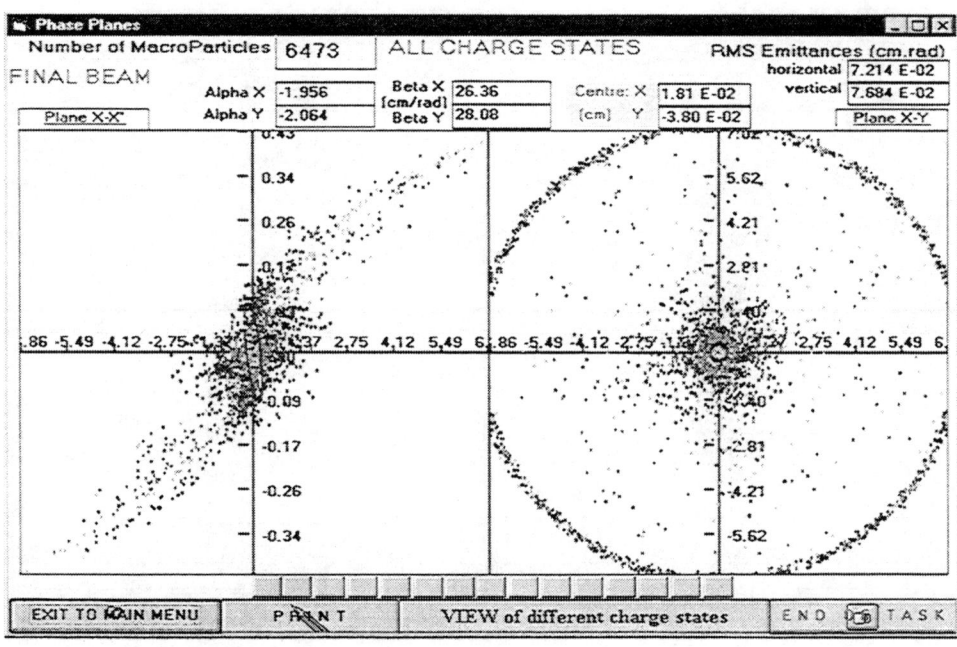

FIGURE 9. Final phase plates of beam emittance and cross density

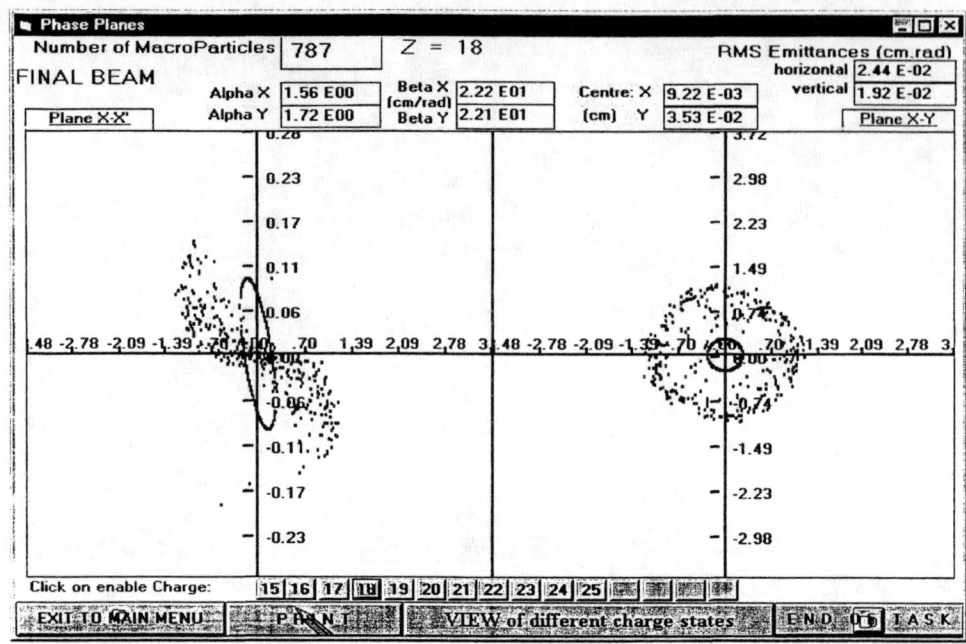

FIGURE 10. Final phase plates of beam emittance and cross density for Ta^{18+} ions

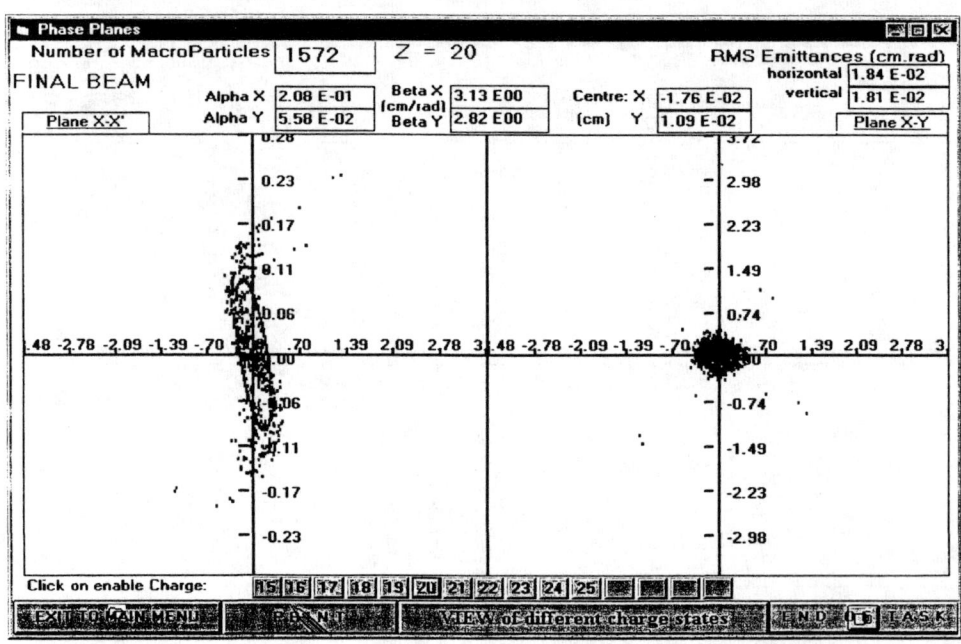

FIGURE 11. Final phase plates of beam emittance and cross density for Ta^{20+} ions

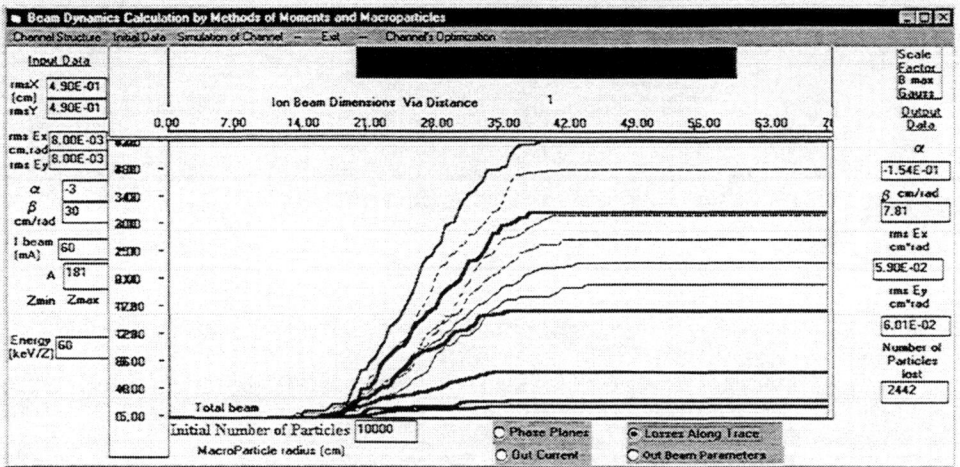

Figure 12. The particle losses along the trace of transportation line

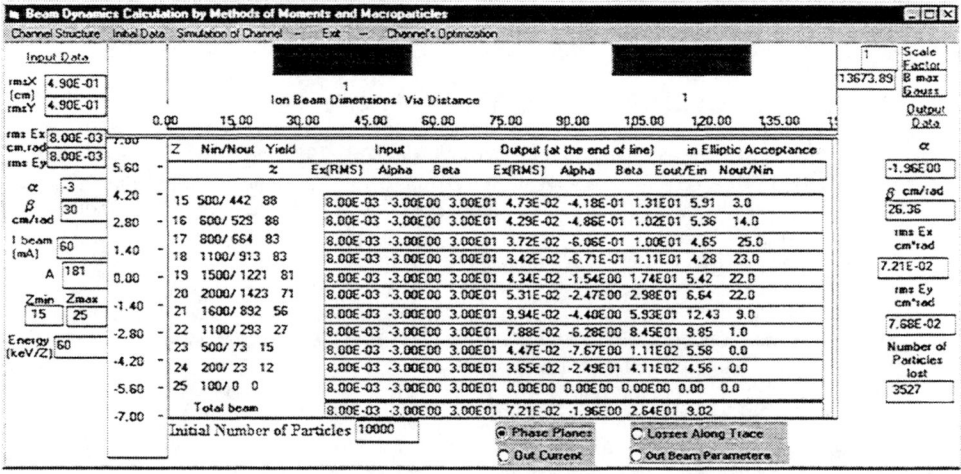

Figure 13 Table of the input and output data for all charge states

CONCLUSION

The program library to calculate transportation lines of intense beam of charged particles is developed. There are two different methods for these calculations. The first of them, based on the method of moments for distribution function allows one to define the mean square root characteristics of the beam in interactive regime. Users can change the parameters of the transportation lines and choose a first approximation of the transportation canal. User interface in this case gives visual results in the screen monitor in the real time of calculations. The detailed characteristics of the beam behavior along the line axes can be calculated with PP-method when this variant of the transportation line is chosen. The program library may be used to calculate the electron and ion beams, the distribution of ion charge states including space charge of the beam.

Parameters of the transportation line and the main beam characteristics can be written in special files and then read again, that provides the convenient conditions for calculations and optimizations of the transportation lines. The results of calculations can be printed and written in files from the program.

REFERENCES

1. Bru B., *GALOPR, a Beam Transport Program, with Space Charge and Bunching*, GANIL Report A88-01.
2. Mittag K. and Sanitz D., Motion - A Versatile Multiparticle Simulation Code, in: *Proceedings of the 1981 Linear Accelerator Conference,* Santa-Fe, New Mexico, October , 1981, Jameson R.A. and Taylor L.S., Eds., LANL Report LA-9234-C, 1982, pp. 156-158,
3. Boicourt G. and Merson J., *PARMILA Users and Reference Manual*, LANL Report LA-UR-90-127, 1990.
4. 5 John W. and Kost C., *Speam a Computor Program for Space Charge Envelopes*, TRIUMF design note TRI-DN-73-11, 1973.
5. Crandall K. R., *Trace 3-D Documentation*, LANL, Report LA-11054-MS, 1987.
6. De Johg M.S. and Heighway E.A., *IEEE Transactions on Nuclear Science NS-30*, 1983, pp. 2666-2668.
7. Farrell J., PATH -A Lumped Element Beam Transport Simulation Program with Space Charge, in: *Proceedings of the Europhysics Conference*, Hahn-Meitrner-Institite fur Kernforschung, Berlin CmBH, W. Busse and R. Zelazny, Eds. (Springer-Verlag, Berlin, New-York, 1984), pp. 267-272.
8. Batygin Y., BEAMPATH: A Program Library for Beam Dynamics simulation in Linear Accelerators, in: *Proceedings of 3d European Particle Accelerator Conference (EPAC92)*, Berlin, 1992, p. 822.
9. Dymnikov A. and Perelstein E., *NIM*, v. 148, p. 567.
10. Kazarinov N.Yu., Perelstein E.A., Shevtsov V.F. *Particle Accelerators*, 1980, v. 10, p. 1.
11. Roshal A.S. *Modelling of Charged Particle Beams*, Atomizdat, Moscow, 1979 (in Russian).
12. Vyalov G.N. et al., *Calculation of High Intensity Injection Line for INR Meson Factory*, (in Russian), JINR Preprint P9-11672, Dubna, 1978.
13. Perelstein E. A. in: *International School for Young Scientists on Problem of Charged particles Accelerators*, JINR, D9-84-817, Dubna, 1984, p. 90.
14. Perelstein E. A. and Shirkov G.D., *Soviet Jornal of Particles & Nuclei*, 18, 1987, p. 64.
15. Sacherer F.J., *IEEE Transections of Nuclear Science*, 1971, NS-18, 3, p.1105.
16. Steffen K., *Basic Course on Accelerators Optics*, CAS CERN 85-19, 1985.
17. Batygin Y., in: Particle Distribution Generator in 4D Phase Space, *Computational Accelerator Physics,* AIP Conference Proceedings 297, Los Alamos 1993, pp. 419-426.

Discussion (Chairman - I. Hofmann, GSI)

I. Hofmann (GSI) I have three items as an introduction to the discussion. One is directly related to heavy driver issues. The other one is on experimental facilities. The third one is actually a spin-off to the other projects. We should not forget, that working in the accelerator field of inertial fusion is not an isolated area, of course. The work we do can be of importance for other projects. And we learn a lot from projects, that people have started in other applications. I am particularly very pleased that people from KEK and JAERI have attended this workshop.

As an example, there were some papers on the electron-ion instability. This work originated from the electron ring accelerator in the seventy's, which was also my own first postdoc assignment. The electron ring accelerator was never built and we don't wish the same for heavy ion fusion. A lot of the work that was done there, was from brilliant people and stood at the forefront of accelerator physics. For instance, the negative mass instability.

The pioneering work of the accelerator side for heavy ion fusion, the Berkeley single beam transport experiment of the early eighty's, has given an enormously important input to the question of stable beam transport below or above ninety degrees phase advance. This has been one of the really significant experiments which - together with the theory - inspired a lot of high current linac design today.

Coming back to the first category: A number of the talks have focused on high phase space density beams. Non-linear space charge compensation by dodecapoles is a very interesting idea. The problem of bends has been largely ignored at the early time of heavy ion fusion, and we just begin to understand the interplay of dispersion, chromaticity and space charge, which is much more important than was thought earlier.

Beam loss is an issue and I bring it up because a place like RIKEN would be ideal to study this subject experimentally. I just want to express my personal feeling, that beam loss is challenging in theory with lots of halo studies which have attracted a number of young people. On the nuclear physics side we really need to get a sound basis for the allowed fractional loss of ions. Everybody knows the importance of nuclear activation by protons of 1 GeV energy. The proton requirements of 1W/m are serious. For heavy ions at heavy ion fusion energies a rough estimate suggests that the activation per particle is up to two orders of magnitude lower. It seems that the heavy ion loss tolerance in terms of W/m is a factor of 1000 better off. A factor of one hundred comes from the lower probability and a factor of ten from the lower average particle current. So the tolerable loss would be one 1 kW/m. That is still a challenging number because the fusion driver is a 50 MW or higher.

R. Davidson (PPPL) What is the status of the experimental data in the 50 MeV/nucleon range?

I. Hofmann (GSI) Not for heavy ions in this energy range. There is a significant material dependence for protons, but if you fragment heavy ions, you produce a large diversity of radioactive elements and that tends to be more independent on material choice.

P.Seidl (LBL) Perhaps we should check the experiments done at the Bevalac several years ago (including NASA's research into long - distance space travel where fragmentation cross section for heavy ions are important). The energy range of these experiments may overlap HIF parameters.

I. Hofmann (GSI) People could ask us eventually why have you not seriously taken this into account so far. I am afraid there was a naive picture around for too long a time, namely that the Coulomb stopping range of heavy ions is so short that nuclear reactions which occur on it are negligible. This is not quite true.

R.Davidson (PPPL) It may be important to understand what the impact is of having heavy ions impact the chamber wall in terms of electrons release from the chamber.

I. Hofmann (GSI) Yes. That has been discussed theoretically but again with almost no experimental data to my knowledge.

P.Zenkevich (ITEP) Also possible secondary emission of heavy ions. It can be dangerous.

I. Hofmann (GSI) Yes, I have noticed that we often deal now with high intensity linacs and a standard question which comes up is what about beam loss in a heavy ion fusion machine? I think we need a complete overview of possibilities to identify the most serious ones.

B.Sharkov (ITEP) I think, that issues really exist and we just neglect them because the energy deposition is really negligible at this energy. For my opinion for the next meeting we can prepare the data of contribution of nuclear reactions.

I. Hofmann (GSI) I know people at GSI have a data up to 10 MeV/nucleon, and from 200 MeV/nucleon to 500 MeV/nucleon, and in between people have extrapolated. At RIKEN I think they could have the medium energy range.

Let us move onto the second category - the experimental facilities where now we have four projects which are the RIKEN (MUSES), TWAC, GSI and induction Integrated Research Experiment (IRE). Either one can be seen as an accelerator test bed for heavy ion drivers, or as a production facility for intense beams to do plasma target experiments. At the second case you worry about the number of particles and kilo Joules. The high kilo Joules with synchrotrons have a lot of appeal. We receive some criticism by other people in our community, why we are studying such an experiment in a synchrotron if the synchrotron is not our choice for a fusion driver ? I guess the people working on TWAC and RIKEN (MUSES) will have to face the same criticism. There is only the US Integrated Research Experiment which picks up the identical concept of the full driver.

Here we have a question to offer for the discussion, namely are we able to predict with existing codes and experience what these machines will actually deliver? To predict for example the incoherent tune shift which one can achieve in a long term storage. The two first projects rely on long term accumulation of ions. So the question is: what tune shift can you achieve ? Can you actually discuss electron cooling of 10^{12} to 10^{13} ions? There is no experience. We just have cooled 10^{10} ions. People have succeeded to cool protons up to 10^{11}. I am sure there are some comments on this issue as we have the real experts in the room.

Another problem is the question of the ion life time in the presence of cooling and in the presence of other effects. I was very pleased to see the new code by the ITEP group that takes into account a very detailed intrabeam scattering life time with other than Gaussian distributions. There are instability issues and electron neutralization, a subject, which was not discussed actually until very recently in this field. I invite you to comment this whole block of issues. May be identify work that needs to be done.

Y.Batygin (RIKEN) My feeling is that there is still no bridge between space charge dominated beam physics in initial part and in storage ring. There is no definition, what is the charge space dominated beams for ring. For initial part we can say that this is a beam where the betatron tune is depressed very serious, almost becomes zero. But for space charge dominated beams in rings there is no even definition. For example, how can we defined, what is ΔQ, from which level we can say, this is a space charge dominated beam? In my opinion it is because phenomena are quite different. Ring itself is very sensitive to resonances, much more sensitive, than initial part. People who study beams dynamics with space charge in rings mostly pay attention on resonances, especially high order resonances. But in the initial part of accelerator facility redistribution due to self space charge forces plays dominant role. Communities are also different and there is no link between them. Some efforts should be done to make this common understanding, how space charge effect transfers from this part to another part.

I. Hofmann (GSI) I have a comment here. Actually there is an interface and this is injection into the ring, now talking about the RF driver. One of the very important conclusion of our HIDIF study was that the need for strong space charge in the linac and also in the final beam line is reduced because we are not able to actually inject a beam into the ring, which has too strong space charge effects. It is a trade off between maximum high phase space density and minimum beam loss. The two things actually work against each other. In the HIDIF study we could only reach a tune shift in the rings after injection of 0.05 whereas, for example, many proton booster synchrotrons reach 0.5. These boosters have a horizontal multiturn injection, and losses are 20-30 percent of the particles at injection. Since we were requiring 1 percent of loss, the possible tune shift in the two-plane injection was only 0.05. For long-tome storage I think it would be also worth looking at the ISR scaling of incoherent tune shift versus storage time. If I remember correctly, if the storage time is of the order of 1 second, they come up with the incoherent tune shift of 0.05. And of course tune shift gets smaller, if you require longer storage time.

P.Zenkevich (ITEP) Some comments. First about tune shift. I would like to underline some contradictions. This high intensity storage rings have not only the problem of current. We would like to have large number of ions, but also we would like to have small momentum spread. At this case there is a danger of appearance of instability. I think, it is necessary to examine several questions. One is a problem of crossing of resonances taking into account space charge. This is a very interesting subject and it is necessary to continue investigation in this field. Second one is a serious contradiction. If you would like to compress the beam, you need a small momentum spread. If we have small momentum spread, we have instability. Point is to examine more accurately the instability in such high current storage ring for making reasonable prediction. Third question is about ion life time and about our code. This is a good idea, because it gives opportunity to take into account all effects.

I think it is possible to include in particle-in-cell code some special programs, which describe the intrabeam scattering, electron cooling, stochastic cooling using Monte-Carlo method. I think, it is necessary to continue the investigation of such type of codes.

D.Koshkarev (ITEP) I want to say some words about heavy ions driver. Now we have two options. One driver is based on resonance linac. Another kind of driver is based on induction linac. I think now the optimum way is to combine this two machines. We can divide the driver for a two parts. From initial energy, from ion source to some intermediate energy, for example for 1 GeV, the resonance RF linac is very convenient. And at the second part, from this energy up to high energy of 10 GeV, the induction linac have many advantages. It is possible to propose combined driver. I think, this is the best way from technical and science point of view.

R.Jameson (LANL) G.Parisi already gave an example this morning on hybrid design. Also 20 years ago we had many workshops on Heavy Ion Fusion in Berkeley and we had even technical proposal on hybrid between resonant front end and induction linac.

I. Hofmann (GSI) Let me make one comment. In our laboratory there is an IH - structure which has been used also for the CERN lead linac. U. Ratzinger who works on this structure is confident that eventually he could design an RF linac based on an IH structure with current as high as 1 or several A with an accelerating gradient up to 6 or 7 MV/m. He eventually would use four beam lines through common tanks.

R.Jameson (LANL) IH structure is pretty common now in linac designs. Also in Russia there was a lot of works previously on IH structure and also on alternating gradient focusing. And especially for heavy ions you can achieve very impressive accelerating gradients.

R.Davidson (PPPL) This is a question for P.Seidl and E.Henestroza, as to whether they could envision a combination of accelerator design in IRE as was suggested.

P.Seidl (LBL) The paths to the several MJ needed at the target have some basic differences in the RF approach and in the induction linac. You can start with a high current, rather low voltage beam and accelerate in an induction accelerator. Or you can start with low current, accelerate up to several GeV with RF and then stack. Then there is the hybrid design mentioned in G.Parisi's talk today. Developing a more integrated technology than this would take some further discussion. At the previous workshop we spoke briefly about our plan to build a high current injector, namely 1.5 or so MeV with beam with current of 0.5 or more Amp. This would be followed by a relatively long AG transport experiment. Concerning the latter, I invite this community to give some thought to the number of lattice periods required to explore some of the key issues that need answers for HIF. The experiment design is based on addressing beam physics and accelerator technology in an AG transport line, where a high current, relatively low emittance beam fills a rather large fraction of the physical aperture.

R.Davidson (PPPL) I would like to mention IRE again. This is a schematic concept of a possible induction linac. Injector source with 84 beams of acceleration in the electrostatic transport system, combine into 21 beams, and the beam interaction and the target chamber region are shown. Parameters are: energy on target 50 - 100 kJ, power density of 50 TW/cm^2, repetition rate of 5 - 10 Hz and average beam power on target is 0.25 - 1 MW. In the view of the heavy ion fusion community in US it is important to minimize the cost of such a facility. These may be generous numbers in the certain sense, but it is in the few hundred million dollars range. So if one is thinking about combining technologies at the IRE stage, it may be too early but is worth thinking about. This is the induction linac parameter range we believe are reasonable operating numbers.

I. Hofmann (GSI) Let me ask our colleges from Russia, if they have any comments on electron cooling. We are constantly asked about this issue: can you apply electron cooling for multi - kJ heavy ions beam for this experimental facility ?

I. Meshkov (JINR) We understand well the physics and principal limits of electron cooling for intensive beams. When we reach in the particle rest frame the ion density comparable with the electron density, we immediately have some qualitative phenomena in interaction of the both beams. For me it is very close to limit of electron cooling application. If we consider parameters of present cooler rings, electron beam density in particle rest frame is of the order of 10^8 electron per cm^3. It well corresponds to beam intensity because typical circumference of rings nowadays is about 100 meters and cooled beam diameter is about half of millimeter. We will get numbers of 10^{10} - 10^{11} and no experience we have with more dense beam. Of course it is very challenging task. It should be studied and this is the next task for people who work with electron cooling. But from the other hand for me it is the only way to have dense multicharge ion beams which we need for fusion business. Stochastic cooling as we know well, is hopeless for this application. And we should think about different possibility how to use electron cooling.

I. Hofmann (GSI) Would you suggest ion particle density or ion charge density to be equal to electron ones?

I. Meshkov (JINR) This is a charge density, because this is a Coulomb interaction.

D. Koshkarev (ITEP) I can say about density problem in storage ring of TWAC. Now we have a slow (1 Hz) repetition rate historically, because we have low amplitude of RF field in the booster ring. But the magnet system have been designed for a very high repetition rate up to 50 Hz. It means we have factor of 40-50 in time. If we will have more powerful RF station, the storage time will be in 50 times shorter and after that the intrabeam scattering will not change our parameters essentially. The electron cooling system will not need in this case.

I. Hofmann (GSI) Are there laser sources at 40 - 50 Hz?

D. Koshkarev (ITEP) This is a problem, but this is a technical problem. So in principal it also can be solved if we have enough money.

I. Hofmann (GSI) Now we will speak about electron neutralization which has been an issue in interesting papers. It was mainly discussed so far for coasting beams. For bunched beams you have realistic hope that there is no chance to accumulate such density to the necessary amount and so we really worry for coasting beam with slow accumulation. But it can be done experimentally, it would be proposed in any machines. People normally try to get away from it.

R.Davidson (PPPL) I would like to understand what would happen if P.Zenkevich's analysis is extended or applied in much higher intensity regimes. Second, what happens if we are able to put a realistic energy spread for electron and ion distributions into a kinetic description. A spread in betatron frequency is able to suppress the instability at moderately high intensity.

P.Zenkevich (ITEP) I would like to underline the dangerous of this instability, but I have a hope in reality that a careful application of clearing electrodes might can help us to suppress it.

I.Hofmann (GSI) The Los Alamos PS has had years of discussions on electron - ion instability and to my knowledge there has been no final convincing conclusion what exactly happened and what really can be done to avoid it.

R.Davidson (PPPL) Just an observation. They don't have clearing electrodes. It is also disappointing that they don't have adequate diagnostics of the electron population and dynamics.

I.Hofmann (GSI) We would have to see if clearing electrode can be used for heavy ion machine with it's low intensity.

R.Jameson (LANL) V. Danilov at ORNL has been working on the PSR instability in terms of space charge effect and resonances and chaos. Actually it is quit interesting argument that PSR instability may in fact come from that things.

R.Davidson (PPPL) On the other hand there is certainly evidence that there is a two-species instability in electron machines: in Cornell, in APS and in ERA. We certainly know there are 2 charge components. Under certain conditions you can generate a strong dipole instability.

I.Hofmann (GSI) It is obviously not an issue for a heavy ion fusion driver, which has to use short pulse so there is no time to build up, but for the experimental facilities it is.

E.Mustafin (ITEP) I have heard from Dieter Mohl at CERN that in LEAR they made experiment to shake out that species by using shaking technique.

I. Meshkov (JINR) What E.Mustafin mentioned is the experiment with clearing of stored ions, ions stored in electron beam, when electron cooling is on, and it was well demonstrated. Now we continue similar experiments with our test bench in Dubna, where we have electron beam and ions stored. In longitudinal magnetic field one can shake them out choosing proper frequency. We have ion mass analyzer developed in our group, and we can choose proper frequency and shake out ions which we like, kind by kind. Such method can be also applied to heavy ion machines.

E.Henestroza (LBL) I have a comment, going back to the question of using IRE as a joint experiment using induction and RF linacs. I think our scale driver does not allow that option. So I do not know if an RF linac can handle such high intensity beams. One ampere beam at 1.6 Megavolt looks like too much space charge for a linac.

E.Henestroza (LBL) Now the comment about predictive capabilities. I think with WARP we have now very good ability to predict most of the beam dynamics features in induction machines. There is still much work to be done to analyze longitudinal instabilities, but I think we can predict most of the beam phenomena in a driver.

D.Koshkarev (ITEP) I agree with you. Resonance linac also can accelerate of course not a high current as induction linac but a good value slightly less than 1 A. When I told about combined RF linac with induction one I mean when we transfer current from RF linac to induction one we need a special system, where the current should be compressed additionally to a high value.

I.Hofmann (GSI) Work I refer to was not a detailed study. When you begin looking for the details, you find out what the problems are. Are there any more comments, even on the last item of spin off to other project ?

I. Meshkov (JINR) There is an overlap with transmutation problem or energy amplifier, because halo problem is also crucial one. And I think the goal task for our community is to communicate with people who are dealing with machines of transmutation, like isochronous cyclotron. Because the beam physics is very close.

I.Hofmann (GSI) This is certainly a very appropriate remark.

AUTHOR INDEX

A

Alekseev, N., 31
Alexandrov, V., 189

B

Batygin, Y. K., 134, 189
Bolshakov, A., 31
Bongardt, K., 159

D

Danilov, S., 8
Deitinghoff, H., 159
Dolbilov, G. V., 85, 99

F

Fateev, A. A., 99
Finley, D., 8
Fujita, K., 108

H

Hanamori, S., 120
Haseroth, H., 1
Homenko, S., 1

I

Ikegami, M., 174

J

Jameson, R. A., 21

K

Katayama, T., 42
Kato, S., 120
Kawata, S., 108, 120
Kazarinov, N., 189
Kikuchi, T., 108, 120
Kondrashev, S., 1
Kugler, H., 1

L

Lisi, N., 1

M

Makarov, K., 1
Mescheryakov, N., 1
Meshkov, I., 42, 65
Mustafin, E., 31

P

Pabst, M., 159
Parisi, G., 159
Petrov, V. A., 99

R

Roerich, V., 1
Rudskoy, I., 1

S

Satov, Y., 1
Scrivens, R., 1
Sery, A., 8
Sharkov, B., 1
Shevtsov, V., 189
Shiltsev, V., 8
Shirkov, G., 189

Shumshurov, A., 1
Sidorov, A. I., 99
Stepanov, A., 1
Syresin, E., 42

Y

Yano, Y., 42
Yazawa, M., 108, 120

T

Takahashi, D., 108

Z

Zenkevich, P. R., 31, 74